The IMA Volumes
in Mathematics
and its Applications

Volume 94

Series Editors
Avner Friedman   Robert Gulliver

Springer
New York
Berlin
Heidelberg
Barcelona
Budapest
Hong Kong
London
Milan
Paris
Santa Clara
Singapore
Tokyo

Institute for Mathematics and its Applications
IMA

The **Institute for Mathematics and its Applications** was established by a grant from the National Science Foundation to the University of Minnesota in 1982. The IMA seeks to encourage the development and study of fresh mathematical concepts and questions of concern to the other sciences by bringing together mathematicians and scientists from diverse fields in an atmosphere that will stimulate discussion and collaboration.

The IMA Volumes are intended to involve the broader scientific community in this process.

Avner Friedman, Director

Robert Gulliver, Associate Director

* * * * * * * * * *

IMA ANNUAL PROGRAMS

Continued at the back

Lorenz T. Biegler   Thomas F. Coleman
Andrew R. Conn   Fadil N. Santosa
Editors

Large-Scale Optimization with Applications

Part III: Molecular Structure and Optimization

With 64 Illustrations

Springer

Lorenz T. Biegler
Chemical Engineering Department
Carnegie Mellon University
Pittsburgh, PA 15213, USA

Thomas F. Coleman
Computer Science Department
Cornell University
Ithaca, NY 14853-0001, USA

Andrew R. Conn
Thomas J. Watson Research Center
P.O. Box 218
Yorktown Heights, NY 10598, USA

Fadil N. Santosa
School of Mathematics
University of Minnesota
Minneapolis, MN 55455, USA

Series Editors:
Avner Friedman
Robert Gulliver
Institute for Mathematics and its
    Applications
University of Minnesota
Minneapolis, MN 55455, USA

Mathematics Subject Classifications (1991): 65Kxx, 90Cxx, 93-XX, 90Bxx, 68Uxx, 92Exx, 92E10, 35R30, 86A22, 73Kxx, 78A40, 78A45

Library of Congress Cataloging-in-Publication Data
Large-scale optimization with applications / Lorenz T. Biegler . . . [et al.].
      p.   cm. — (The IMA volumes in mathematics and its applications ; 92–94)
      Presentations from the IMA summer program held July 10–28, 1995.
      Includes bibliographical references.
      Contents: pt. 1. Optimization in inverse problems and design — pt. 2. Optimal design and control — pt. 3. Molecular structure and optimization.
      ISBN 0-387-98286-8 (pt. 1 : alk. paper). — ISBN 0-387-98287-6 (pt. 2 : alk. paper). — ISBN 0-387-98288-4 (pt. 3 : alk. paper)
      1. Mathematical optimization—Congresses.   2. Programming (Mathematics)—Congresses.   3. Inverse problems (Differential equations)—Congresses.   4. Engineering design—Congresses.
      5. Molecular structure—Mathematical models—Congresses.
      I. Biegler, Lorenz T.   II. Series: IMA volumes in mathematics and its applications ; v. 92–94.
      QA402.5.L356   1997
      500.2′01′5193—dc21   97-22879

Printed on acid-free paper.

Production managed by Allan Abrams; manufacturing supervised by Johanna Tschebull.
Camera-ready copy prepared by the IMA.
Printed and bound by Braun-Brumfield, Inc., Ann Arbor, MI.
Printed in the United States of America.

9 8 7 6 5 4 3 2 1

ISBN 0-387-98288-4 Springer-Verlag New York Berlin Heidelberg   SPIN 10632914

FOREWORD

This IMA Volume in Mathematics and its Applications

LARGE-SCALE OPTIMIZATION WITH APPLICATIONS, PART III: MOLECULAR STRUCTURE AND OPTIMIZATION

is one of the three volumes based on the proceedings of the 1995 IMA three-week Summer Program on "Large-Scale Optimization with Applications to Inverse Problems, Optimal Control and Design, and Molecular and Structural Optimization." The other two related proceedings appeared as Volume 92: Large-Scale Optimization with Applications, Part I: Optimization in Inverse Problems and Design and Volume 93: Large-Scale Optimization with Applications, Part II: Optimal Design and Control.

We would like to thank Lorenz T. Biegler, Thomas F. Coleman, Andrew R. Conn, and Fadil N. Santosa for their excellent work as organizers of the meetings and for editing the proceedings.

We also take this opportunity to thank the National Science Foundation (NSF), the Department of Energy (DOE), and the Alfred P. Sloan Foundation, whose financial support made the workshops possible.

Avner Friedman

Robert Gulliver

GENERAL PREFACE FOR
LARGE-SCALE OPTIMIZATION
WITH APPLICATIONS, PARTS I, II, AND III

There has been enormous progress in large-scale optimization in the past decade. In addition, the solutions to large nonlinear problems on moderate workstations in a reasonable amount of time are currently quite possible. In practice for many applications one is often only seeking improvement rather than assured optimality (a reason why local solutions often suffice). This fact makes problems that at first sight seem impossible quite tractable. Unfortunately and inevitably most practitioners are unaware of some of the most important recent advances. By the same token, most mathematical programmers have only a passing knowledge of the issues that regularly arise in the applications.

It can still be fairly said that the vast majority of large-scale optimization modeling that is carried out today is based on linearization, undoubtedly because linear programming is well understood and known to be effective for very large instances. However, the world is not linear and accurate modeling of physical and scientific phenomena frequently leads to large-scale nonlinear optimization.

A three-week workshop on Large-Scale Optimization was held at the IMA from July 10 to July 28, 1995 as part of its summer program. These workshops brought together some of the world's leading experts in the areas of optimization, inverse problems, optimal design, optimal control and molecular structures. The content of these volumes represent a majority of the presentations at the three workshops. The presentations, and the subsequent articles published here are intended to be useful and accessible to both the mathematical programmers and those working in the applications. Perhaps somewhat optimistically, the hope is that the workshops and the proceedings will also initiate some long-term research projects and impart to new researchers the excitement, vitality and importance of this kind of cooperation to the applications and to applied mathematics.

The format of the meetings was such that we tried to have an invited speaker with expertise in an application of large-scale optimization describe the problem characteristics in the application, current solution approaches and the difficulties that suggest areas for future research. These presentations were complemented by an optimization researcher whose object was to present recent advances related to the difficulties associated with the topic (*e.g.*, improved methods for nonlinear optimization, global optimization, exploiting structure). One difficulty was that although it is possible (but perhaps not desirable) to isolate a particular application, the optimization methods tended to be intertwined in all of the topics.

These Proceedings include the same mix of details of the application, overview of the optimization techniques available, general discussions of the difficulties and areas for future research.

We are grateful to all the help we had from the IMA, and in particular we would like to single out Avner Friedman, Robert Gulliver and Patricia Brick whose help and support was invaluable. Patricia Brick is especially acknowledged for all of her efforts typesetting and assembling these volumes. The speakers, the attendees and the diligent reviewers of the submitted papers also deserve our acknowledgment; after all, without them there would be no proceedings. Finally we would like to thank those agencies whose financial support made the meeting possible: The National Science Foundation, the Department of Energy, and the Alfred P. Sloan Foundation.

Lorenz T. Biegler

Thomas F. Coleman

Andrew R. Conn

Fadil N. Santosa

PREFACE FOR PART III

Many important molecular conformation problems, such as protein folding, are expressed as global minimization problems. It is the fact that local minimization is insufficient, that markedly differentiates this volume from the previous two.

Unfortunately, global minimization problems that result from models of molecular conformation are usually intractable. For example, simple 1-dimensional versions of distance conformation problems are NP-hard. Nevertheless, there has been significant recent progress in the design of promising heuristic strategies (often involving the use of high-performance parallel computers) for computing approximate global minimizers. The purpose of the sessions represented in this volume was to discuss the new algorithmic advances for global minimization in the context of protein folding and related molecular minimization problems. Emphasis was on practical shortcomings of current approaches, outstanding problems and questions, and the use of high-performance parallel computers.

Lorenz T. Biegler

Thomas F. Coleman

Andrew R. Conn

Fadil N. Santosa

CONTENTS

Large-Scale Optimization with Applications,
Part III: Molecular Structure and Optimization

CONTENTS OF PART I: OPTIMIZATION IN INVERSE PROBLEMS AND DESIGN

CONTENTS OF PART II: OPTIMAL DESIGN AND CONTROL

CGU: AN ALGORITHM FOR MOLECULAR STRUCTURE PREDICTION

K.A. DILL[*], A.T. PHILLIPS[†], AND J.B. ROSEN[‡]

Abstract. A global optimization method is presented for predicting the minimum energy structure of small protein-like molecules. This method begins by collecting a large number of molecular conformations, each obtained by finding a local minimum of a potential energy function from a random starting point. The information from these conformers is then used to form a convex quadratic global underestimating function for the potential energy of all known conformers. This underestimator is an L_1 approximation to all known local minima, and is obtained by a linear programming formulation and solution. The minimum of this underestimator is used to predict the global minimum for the function, allowing a localized conformer search to be performed based on the predicted minimum. The new set of conformers generated by the localized search serves as the basis for another quadratic underestimation step in an iterative algorithm. This algorithm has been used to predict the minimum energy structures of heteropolymers with as many as 48 residues, and can be applied to a variety of molecular models. The results obtained also show the dependence of the native conformation on the sequence of hydrophobic and polar residues.

AMS(MOS) subject classifications. Primary 65K05: Secondary 90C26, 90C90.

1. Introduction. It is widely accepted that the folded state of a protein is completely dependent on the one-dimensional linear sequence (i.e. "primary" sequence) of amino acids from which it is constructed: external factors, such as helper (chaperone) proteins, present at the time of folding have no effect on the final, or native, state of the protein. Furthermore, the existence of a unique native conformation, in which residues distant in sequence but close in proximity exhibit a densely packed hydrophobic core, suggests that this 3-dimensional structure is largely encoded within the sequential arrangement of these hydrophobic (H) and polar (P) amino acids. The assumption that hydrophobic interaction is the single most dominant force in the correct folding of a protein also suggests that simplified potential energy functions, for which the terms involve only pairwise H-H attraction and steric overlap repulsion, may be sufficient to guide computational search strategies to the global minimum representing the native state.

During the past 20 years, a number of computer algorithms have been developed that aim to predict the fold of a protein (see for example [3], [5], [8], [10]). Such approaches are generally based on two assumptions. First, that there exists a potential energy function for the protein; and

---

[*] Department of Pharmaceutical Chemistry, University of California San Francisco, San Francisco, CA 94118.

[†] Computer Science Department, United States Naval Academy, Annapolis, MD 21402.

[‡] Computer Science and Engineering Department, University of California San Diego, San Diego, CA 92093.

second that the folded state corresponds to the structure with the lowest
potential energy (minimum of the potential energy function) and is thus
in a state of thermodynamic equilibrium. This view is supported by in
vitro observations that proteins can successfully refold from a variety of
denatured states.

2. A simple polypeptide model. Computational search methods
are not yet fast enough to find global optima in real-space representations
using accurate all-atom models and potential functions. A practical conformational search strategy will require both a simplified molecular model
with an associated potential energy function which consists of the dominant
forces involved in protein folding, and also a global optimization method
which takes full advantage of any special properties of this kind of energy
function. In what follows, we describe a global optimization algorithm
which has been successfully used for one such simplified model. We then
describe a more realistic model, which we believe will permit the use of our
algorithm on small protein molecules.

In our initial testing of the CGU algorithm (to be described shortly), we
chose to use a simple "string of beads" model consisting of n monomers C
connected in sequence (see Figure 2.1). The 3-dimensional position of each
monomer, relative to the previous monomers in the sequence, is defined
by the parameters l (the "bond lengths"), θ (the "bond angles"), and ξ
(the backbone "dihedral angles"). Of these we have chosen to fix l and
θ (the reasons for this will become clear later), thus reducing the number
of independent parameters necessary to uniquely define a 3-dimensional
conformation to only $n - 1$. In order to model the H-P effects that are
encoded within the backbone sequence, each "bead" C in this simplified
model is categorized as either hydrophobic (H) or polar (P).

Corresponding to this simplified polypeptide model is a potential energy function also characterized by its simplicity. This function includes
just three components: a contact energy term favoring pairwise $H - H$
residues, a steric repulsive term which rejects any conformation that would
permit unreasonably small interatomic distances, and a main chain torsional term that allows only certain preset values for the backbone dihedral
angles ξ. Despite its simplicity, the use of this type of potential function
has already proven successful in studies conducted independently by Sun,
Thomas, and Dill [13] and by Srinivasan and Rose [11]. Both groups have
demonstrated that this type of potential function is sufficient to accurately
model the forces which are most responsible for folding proteins. Here our
energy function is somewhat different from either of those. The specific
potential function used initially in this study has the following form:

$$(2.1) \qquad E_{\text{total}} = E_{ex} + E_{hp} + E_\xi$$

where the steric repulsion and hydrophobic attraction terms $E_{ex} + E_{hp}$
can conveniently be combined and represented by the well known Lennard-

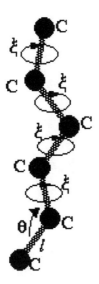

FIG. 2.1. *Simplified "String of Beads" Polypeptide Model.*

Jones pairwise potential function

$$\sum_{|i-j|>2} \epsilon_{ij} \left(\left(\frac{\sigma_{ij}}{r_{ij}} \right)^{12} - 2H_{ij} \left(\frac{\sigma_{ij}}{r_{ij}} \right)^{6} \right).$$

This term defines the potential energy contributions of all beads separated by more than two along the primary chain. The Lennard-Jones coefficients ϵ_{ij} and σ_{ij} are constants defined by the relationships between the two specific beads (e.g. amino acids) involved. The terms involving r_{ij} in the Lennard-Jones expression represent the Euclidean distances between beads i and j. The constant $H_{ij} = 1$ if beads i and j are both H-type (attractive monomers), and hence both a repulsive force (ensuring that the chain is "self-avoiding") and an attractive force (since the beads are H-H) are added to the potential energy (see Figure 2.2). On the other hand, $H_{ij} = 0$ if the beads i and j are H-P, P-H, or P-P pairs, so that the Lennard-Jones contribution to the total potential energy is just the repulsive force that ensures self-avoidance.

A trigonometric based penalty implementing the potential energy term E_ξ in equation 1 was used in these tests, and had the following form:

$$E_\xi = \sum_i C_1 (1 + \cos(3\xi_i)).$$

Using this term, there are only three "preferred" backbone dihedral angles of $60°$, $180°$, and $300°$ with all others penalized to some extent (determined

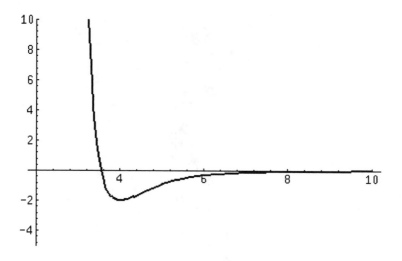

FIG. 2.2. *Lennard-Jones Pair Potential with* $H_{ij} = 1$, $\epsilon_{ij} = 2$, *and* $\sigma_{ij} = 4$.

by the constant $C1$). The purpose of this term is to mimic, in some elementary sense, the restrictive nature of the Ramachandran plot (see [4]) for each residue in a realistic protein model.

3. The CGU global optimization algorithm. One practical means for finding the global minimum of the polypeptide's potential energy function is to use a global underestimator to localize the search in the region of the global minimum. This CGU (convex global underestimator) method is designed to fit all known local minima with a convex function which underestimates all of them, but which differs from them by the minimum possible amount in the discrete L_1 norm (see Figure 3.1). Any non-negative linear combination of convex functions can be used for the underestimator, but for simplicity we use convex quadratic functions. The minimum of this underestimator is used to predict the global minimum for the function, allowing a localized conformer search to be performed based on the predicted minimum. A new set of conformers generated by the localized search then serves as a basis for another quadratic underestimation. After several repetitions, the global minimum can be found with reasonable assurance.

This method, first described in [9], is presented in terms of the differentiable potential energy function $E_{\text{total}}(\phi)$, where $\phi \in \mathbf{R}^{n-1}$ (n represents the number of residues in the polypeptide chain), and where $E_{\text{total}}(\phi)$ has many local minima. Thus, ϕ is a vector of $n-1$ backbone dihedral angles. Defining $\tau = n - 1$, then to begin the iterative process, a set of $k \geq 2\tau + 1$ distinct local minima are computed. This can be done with relative ease by using an efficient unconstrained minimizer, starting with a large enough set of points chosen at random in an initial hyperrectangle $H\phi$, which is

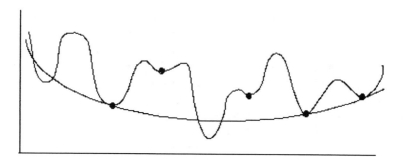

FIG. 3.1. *The Convex Global Underestimator (CGU).*

assumed to enclose the entire torsion angle space.

Assuming that $k \geq 2\tau + 1$ local minima $\phi^{(j)}$, for $j = 1, \ldots, k$, have been computed, a convex quadratic underestimator function $F(\phi)$ is now fitted to these local minima so that it underestimates all the local minima, and normally interpolates $E_{\text{total}}(\phi^{(j)})$ at $2\tau + 1$ points (see Figure 3.1). This is accomplished by determining the coefficients in the function $F(\phi)$ so that

$$(3.1) \qquad \delta_j = E_{\text{total}}(\phi^{(j)}) - F(\phi^{(j)}) \geq 0$$

for $j = 1, \ldots, k$, and where $\sum_{j=1}^{k} \delta_j$ is minimized. That is, the difference between $F(\phi)$ and $E_{\text{total}}(\phi)$ is minimized in the discrete L_1 norm over the set of k local minima $\phi^{(j)}$, $j = 1, \ldots, k$. Although many choices are possible, the underestimating function $F(\phi)$ selected for the CGU method is a separable convex quadratic given by

$$(3.2) \qquad F(\phi) = c_0 + \sum_{i=1}^{\tau} \left(c_i \phi_i + \frac{1}{2} d_i \phi_i^2 \right).$$

Note that c_i and d_i appear linearly in the constraints of equation 3.1 for each local minimum $\phi^{(j)}$. Convexity of this quadratic function is guaranteed by requiring that $d_i \geq 0$ for $i = 1, \ldots, \tau$.

Additionally, in order to guarantee that $F(\phi)$ attains its global minimum $F_{\min}$ in the hyperrectangle $H\phi = \{\phi_i : 0 \leq \underline{\phi}_i \leq \phi_i \leq \overline{\phi}_i \leq 2\pi\}$, the following additional set of constraints are imposed on the coefficients of $F(\phi)$:

$$(3.3) \qquad c_i + \underline{\phi}_i d_i \leq 0 \text{ and } c_i + \overline{\phi}_i d_i \geq 0 \text{ for } i = 1, \ldots, \tau.$$

Note that the satisfaction of equation (3.3) implies that $c_i \leq 0$ and $d_i \geq 0$ for $i = 1, \ldots, \tau$.

The unknown coefficients c_i, $i = 0, \ldots, \tau$, and d_i, $i = 1, \ldots, \tau$, can be determined by a simple linear programming formulation and solution, and since the convex quadratic function $F(\phi)$ gives a global approximation to the local minima of $E_{\text{total}}(\phi)$, then its easily computed global minimum function value $F_{\min}$ is a good candidate for an approximation to the global minimum of the potential energy function $E_{\text{total}}(\phi)$. The complete details of this linear programming formulation are given in [9], and so are not presented here.

The convex quadratic underestimating function $F(\phi)$ determined by the values $c \in \mathbf{R}^{\tau+1}$ and $d \in \mathbf{R}^{\tau}$ provide a global approximation to the local minima of $E_{\text{total}}(\phi)$, and its easily computed global minimum point $F_{\min}$ is given by $(\phi_{\min})_i = -c_i/d_i$, $i = 1, \ldots, \tau$, with corresponding function value $F_{\min}$ given by $F_{\min} = c_0 - \sum_{i=1}^{\tau} c_i^2/(2d_i)$. The value $F_{\min}$ is a good candidate for an approximation to the global minimum of the potential energy function $E_{\text{total}}(\phi)$, and so $\phi_{\min}$ can be used as an initial starting point around which additional configurations (i.e. local minima) should be generated. These local minima are added to the set of all known local minima, and the process is repeated. Before each iteration of this process, it is necessary to reduce the volume of the hyperrectangle $H\phi$ over which the new configurations are produced so that a tighter fit of $F(\phi)$ to the local minima "near" $\phi_{\min}$ is constructed.

If E_c is a cutoff energy, then one means for modifying the size of the hyperrectangle $H\phi$ at any step is to let $H\phi = \{\phi : F(\phi) \leq E_c\}$. Clearly, if E_c is reduced, the size of $H\phi$ is also reduced. At every iteration the predicted global minimum value $F_{\min}$ satisfies $F_{\min} \leq E_{\text{total}}(\phi^*)$, where ϕ^* is the smallest *known* local minimum conformation computed so far (see Figure 3.2). Therefore, $E_c = E_{\text{total}}(\phi^*)$ is often a good choice. If at least one improved point ϕ, with $E_{\text{total}}(\phi) < E_{\text{total}}(\phi^*)$, is obtained in each iteration, then the search domain $H\phi$ will strictly decrease at each iteration, and may decrease substantially in some iterations (see Figure 3.3). Such a means for reducing the search domain $H\phi$ does not of course guarantee that the true global minimum will be found. In fact, it should be clear that the true global solution may be excluded from the new search domain if it avoids detection as a local minimum solution at every iteration. Hence it is very important that the initial set of k distinct local minima be sufficiently large so that either the true global minimum is included among them, or so that the global convex underestimator $F(\phi)$ accurately models and predicts the global structure of $E_{\text{total}}(\phi)$. As a general rule of thumb (based only on computational experience), $k = 10(2\tau + 1)$ is sufficient for this purpose.

Based on the general discussion above and the details provided in [9], the CGU algorithm can be succinctly described as follows:

    1. Compute $k \geq 2\tau + 1$ distinct local minima $\phi^{(j)}$, for $j = 1, \ldots, k$, of the function $E_{\text{total}}(\phi)$.

2. Compute the convex quadratic underestimator function

$$F(\pi) = c_0 + \sum_{i=1}^{\tau} \left(c_i \phi_i + \frac{1}{2} d_i \phi_i^2 \right)$$

by solving a linear program (see [9] for details). The optimal solution to this linear program directly provides the values of c and d.

3. Compute the predicted global minimum point $\phi_{\min}$ given by $(\phi_{\min})_i = -c_i/d_i$, $i = 1, \ldots, \tau$ with corresponding function value $F_{\min}$ give by $F_{\min} = c_0 - \sum_{i=1}^{\tau} c_i^2/(2d_i)$.

4. If $\phi_{\min} = \phi^*$, where $\phi^* = \arg\min\{E_{\text{total}}(\phi^{(j)}), j = 1, 2, \ldots\}$ is the best local minimum found so far, then stop and report ϕ^* as the approximate global minimum conformation.

5. Reduce the volume of the hyperrectangle $H\phi$ over which the new configurations will be produced by using the rule $H\phi = \{\phi : F(\phi) \le E_c\}$ where $E_c = E_{\text{total}}(\phi^*)$.

6. Use $\phi_{\min}$ as an initial starting point around which additional local minima $\phi^{(j)}$ of $E_{\text{total}}(\phi)$ (restricted to the hyperrectangle $H\phi$) are generated.

7. Return to step 2.

While the number of new local minima to be generated in step 6 is unspecified, a value exceeding $2\tau + 1$ would of course be required for the construction of another convex quadratic underestimator in the next iteration (step 2). In the computational tests presented in the next section, we have chosen to use $10(2\tau + 1)$ starting points for both steps 1 and 6 in an attempt to generate at least $2\tau + 1$ distinct local minima.

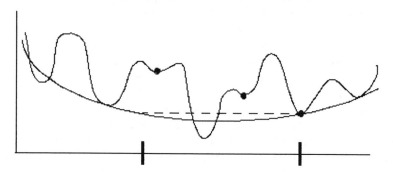

FIG. 3.2. *Defining the Hyperrectangle $H\phi$.*

The rate and method by which the hyperrectangle size is modified are important in determining the convergence properties of the CGU algorithm. It can be shown that if the convex underestimator $F(\phi)$ does in fact underestimate the global minimum of $E_{\text{total}}(\phi)$ at every iteration of

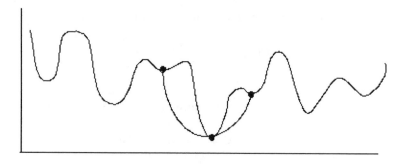

FIG. 3.3. *The New CGU Over a Reduced Hyperrectangle H φ.*

the CGU algorithm, then by appropriately applying "branch-and-bound" techniques to this method, finite convergence to the global minimum can be guaranteed (albeit in a possibly exponential number of steps). Note that the underestimator need not actually underestimate *all* local minima for this property to be true, it only need underestimate the global solution at each step. Since $F(\phi)$ underestimates all known local minima in the current hyperrectangle $H\phi$ (by construction), it is very likely that it will also underestimate the global minimum. While this is not guaranteed to be the case, our computational experience shows that is usually satisfied. Another method of constructing a convex global underestimator, and a related global optimization algorithm, is described elsewhere in [6], but no computational comparison of these two methods has yet been made.

4. Computational results for the simplified model. The computational results presented below for the simplified model were obtained on a network of eight Sun SparcStations using the MPI message passing system for communication between the CPUs. Steps 1 and 6 of the algorithm (presented in section 3) were performed in parallel on all eight of the processors, while the remaining steps of the algorithm were done sequentially on a single designated "master" processor.

A detailed set of computational results for the CGU method have previously been presented in [9]. In that paper, the method was tested on a large sample of homopolymer sequences (that is, all residues are hydrophobic) ranging in size from only 4 residues ("beads") to as many as 30. This paper presents only two additional test cases, but these tests serve to demonstrate that the CGU method can be successfully applied to larger heteropolymer sequences (i.e. mixed sequences of H and P). The two HP sequences tested were designed by E. Shakhnovich as part of a friendly competition between his group at Harvard and the Dill group at the University of California, San Francisco. In that competition, the Harvard group designed a set of 3-dimensional lattice-based 48-mer HP sequences with a known folded target structure (also restricted to the lattice) which they denoted "putative native state" (PNS). The PNS was not known to be the global solution, since it was computed by an inverse folding technique using a Monte Carlo method. The object of that competition was to see if the Dill group could find the PNS (or a folded state with an even lower energy) using their own algorithms, but given only the primary HP sequence. Ten HP sequences were tested, and of these we have selected two representative ones, the sequences labeled #8 and #10 (see [14]).

The computational tests presented below serve to illustrate two points: (1) that the CGU method can be practical for moderate size sequences (in this case 48-mer sequences), (2) and that the global minimum energy is in fact *very highly dependent* on the primary sequence.

When applied to the first of these sequences, sequence #8, the CGU method found the folded state, with an associated minimum energy of -87.57, as shown in Figure 4.1 (the dark grey beads are P type while the light grey beads are H type). For this folded conformation (which we will denote $F\#8$), if the sequence had consisted of all P type monomers, then corresponding energy for $F\#8$ would be $+128.99$, whereas if it had consisted of all H type monomers, the energy would have been -239.97. Furthermore, when the sequence is fully extended, i.e. all backbone dihedrals ξ are set to $180°$, then the corresponding energy for sequence #8 is -4.22. This may also be considered an upper bound.

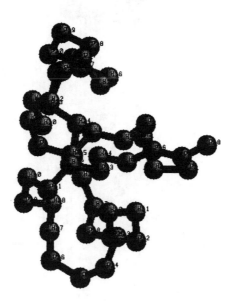

FIG. 4.1. *Folded State ($F\#8$) for HP Sequence #8.*

Sequence #8 is PHHPHHHPHHHHPPHHHHPPPPPPHPHHPPHHPHPPPHHPHPHPHHPPP

Recall that $F\#8$ is the presumed global minimum conformation for the HP sequence #8. If this sequence is replaced by all H or all P type monomers, then $F\#8$ is not necessarily even a local minimum. Figures 4.2 and 4.3 show the conformations which result from a single local minimization (i.e. relaxation) beginning from state $F\#8$ with these two homopolymer sequences in place of sequence #8. Clearly, they are decidedly different. Table 4.1 summarizes these various results.

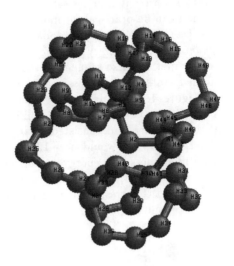

FIG. 4.2. *Relaxation from F#8 Using 48-mer All H-type Homopolymer.*

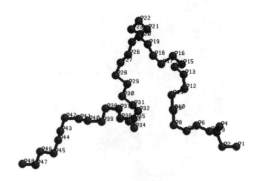

FIG. 4.3. *Relaxation From F#8 Using 48-mer All P-type Homopolymer.*

Sequence #10 is PHHPPPPPPHHPPPHHHPHPPHPHHPPHPPHPPHHPPHHHHHHHPPHH

TABLE 4.1
Dependence of Energy on Primary Sequence (Based on # 8)

	All H-type Homopolymer	HP Sequence #8	All P-type Homopolymer
Fully Extended	-19.55	-4.22	$+1.27$
$F\#8$	-239.97	-87.57	$+128.99$
Relaxation from $F\#8$	-334.22	-87.57	$+13.06$

TABLE 4.2
Dependence of Energy on Primary Sequence (Based on # 10)

	All H-type Homopolymer	HP Sequence #10	All P-type Homopolymer
Fully Extended	-19.55	-3.98	$+1.27$
$F\#10$	-243.87	-97.22	$+133.91$
Relaxation from $F\#10$	-315.41	-97.22	$+10.56$

A similar analysis was performed for sequence #10. Figure 4.4 shows the folded conformation (denoted $F\#10$) for sequence #10, which has a corresponding energy value of -97.22. Figures 4.5 and 4.6 show the relaxed conformations obtained when sequence #10 is replaced by an all H and an all P homopolymer sequence, and Table 4.2 summarizes the results according to energies for each state. Like sequence #8, sequence #10 consists of $50H$ and $50F\#10$ are decidedly different (compare Figures 4.1 and 4.4). In fact, Table 4.3 shows that the energy obtained by sequence #8 when placed into state $F\#10$ (the "native" state for sequence #10) is actually considerably above its minimum energy in $F\#8$ (compare -23.20 to -87.57). Likewise, when sequence #10 is evaluated in state $F\#8$ it also obtains a much higher energy (compare $+39.48$ to -97.22). Furthermore, even if these "non-native" conformations are allowed to relax to a local minimum, Table 4.3 clearly shows that the result remains non-native. Hence, the global minimum energy and associated conformation of an HP sequence is very highly dependent on the primary sequence used. Figures 4.7 and 4.8 show the two "relaxed" conformations obtained by locally minimizing each sequence from the "other" sequences native state.

It is also the case that removing the lattice restriction, as we have done, gives a very different native conformation with the identical HP sequence. This is seen by comparing the conformations in Figures 4.1 and 4.4 with those presented in [14]. However, it should also be noted that the energy

TABLE 4.3
Comparison of Energies for Sequences #8 and #10

	Sequence #8	Sequence #10
F#8	−87.57	+39.48
F#10	−23.20	−97.22
Relaxation from F#8	−87.57	−33.90
Relaxation from F#10	−54.28	−97.22

function used in these tests, equation (2.1), is not related to the lattice-based energy function that was used in the "competition". Hence, one should not expect the lattice-based results to provide a reasonable approximation to the "relaxed" 3-dimensional folded states of a molecules when not restricted to a lattice.

FIG. 4.4. *Folded State (F#10) for HP Sequence #10.*

FIG. 4.5. *Relaxation From F#10 Using 48-mer All H-type Homopolymer.*

FIG. 4.6. *Relaxation From F#10 using 48-mer All P-type Homopolymer.*

FIG. 4.7. *Conformation Obtained by Relaxation of Sequence #8 from State F#10.*

FIG. 4.8. *Conformation Obtained by Relaxation of Sequence #10 from State F#8.*

5. A more detailed polypeptide model. As previously stated, by using a simplified polypeptide model, the complexity of the problem formulation can be reduced to an acceptable level for optimization techniques. Unfortunately though, the simplifications made in section 2 do not provide a very realistic model of true protein sequences. The simplifications were made only to illustrate and test the applicability of the CGU global optimization algorithm to protein structure prediction. Hence, for the CGU algorithm to be a practical method for determining tertiary structure, it must be applied to a more detailed and realistic polypeptide model.

In real proteins, each residue in the primary sequence is characterized by its backbone components NH-C_αH-C'O and one of 20 possible amino acid sidechains attached to the central C_α atom. A key element of this more detailed model is that each *sidechain* is classified as either hydrophobic or polar, and is represented by only a single "virtual" center of mass atom. Thus the potential energy function again involves only three terms: excluded volume repulsive forces between all pairs of atoms, a very powerful attractive force between each pair of hydrophobic sidechain center of mass atoms, and a torsional penalty disallowing conformations which do not exist. Since the residues in this model come in only two forms, H (hydrophobic) and P (polar), where the H-type monomers exhibit a strong pairwise attraction, the lowest free energy state is obtained by those conformations with the greatest number of H-H "contacts" (see [1], [12]). One significant advantage of this detailed formulation of the folding problem is that it allows the model to take advantage of known scientific knowledge about the chemical structure of real sequences of molecules. The use of knowledge such as the Ramachandran plot (see [4]), which specifies the allowable angles between consecutive amino acids in proteins, also greatly simplifies the problem.

Realistic molecular structure information is often given in terms of internal molecular coordinates which consist of bond lengths l (defined by every pair of consecutive backbone atoms), bond angles ϕ (defined by every three consecutive backbone atoms), and the backbone dihedral angles φ, ψ, and ω where φ gives the position of C' relative to the previous three consecutive backbone atoms C'-N-C_α, ψ gives the position of N relative to the previous three consecutive backbone atoms N-C_α-C', and ω gives the position of C_α relative to the previous three consecutive backbone atoms C_α-C'-N. Hence, the backbone of a protein consisting of n amino acid residues can be completely represented in 3-dimensional space using these parameters, as shown in Figure 5.1.

Fortunately, these $9n-6$ parameters (for an n-residue structure) do not all vary independently. In fact, some of these ($7n-4$ of them, to be precise) are regarded as fixed since they are found to vary within only a very small neighborhood of an experimentally determined value. Among these are the $3n-1$ backbone bond lengths l between the pairs of consecutive atoms N-C' (fixed at 1.32 Å), C'-Ca (fixed at 1.53 Å), and C_α-N (fixed at 1.47

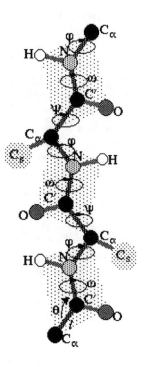

FIG. 5.1. *More Detailed Polypeptide Model.*

$\mathring{A}$). Also, the $3n-2$ backbone bond angles θ defined by N-C_α-C$'$ (110°), C_α-C$'$-N (114°), and C$'$-N-C_α (123°) are also fixed at their ideal values. It is for these reasons that l and θ were also fixed in the simplified model in section 2. Finally, the $n-1$ peptide bond dihedral angles ω are fixed in the trans (180°) conformation. This leaves only the $n-1$ backbone dihedral angle pairs (φ, ψ) in the reduced representation model. These also are not completely independent; in fact, they are severely constrained by known chemical data (the Ramachandran plot) for each of the 20 amino acid residues.

Furthermore, since the atoms from one C_α to the next C_α along the backbone can be grouped into rigid *planar* peptide units, there are no extra parameters required to express the 3-dimensional position of the attached O and H peptide atoms. These bond lengths and bond angles are also known and fixed at 1.24 $\mathring{A}$ and 121° for O, and 1.0 $\mathring{A}$ and 123° for H. Likewise, since each sidechain is represented by only a single center of mass "virtual atom" C_s, no extra parameters are needed to define the position of each sidechain with respect to the backbone mainchain. The following table (Table 5.1) of sidechain bond lengths (between the backbone atom C_α and the sidechain center of mass atom C_s), sidechain bond angles (formed by the sequence

N-$C_\alpha C_s$), and sidechain torsion angles (between Cs and the plane formed by the backbone sequence N-C_α-C′) were used to fix the position of each sidechain atom. The twenty amino acids are also classified into two groups (shown in the table), hydrophobic and polar, according to the scale given by Miyazawa and Jernigan in [7].

Corresponding to this new more detailed polypeptide model is a new potential energy function. As in the simplified model of section 2, this function includes the three components in equation (2.1): a contact energy term E_{hp} favoring pairwise H-H residues, a steric repulsive term E_{ex} which rejects any conformation that would permit unreasonably small interatomic distances, and a main chain torsional term $E_{\varphi\psi}$ (replacing E_ξ in the simplified model) that allows only those (φ, ψ) pairs which are permitted by the Ramachandran plot. The specific potential function used in this more detailed and accurate polypeptide model is most similar to the Sun/Thomas/Dill [13] potential function, which, as stated earlier, has already been proven successful in studies conducted independently by Sun, Thomas, and Dill and by Srinivasan and Rose [11]. In particular, the excluded volume energy term E_{ex} and the hydrophobic interaction energy term E_{hp} are defined in this case as follows:

$$E_{ex} = \sum_{ij} \frac{C_1}{1.0 + \exp((d_{ij} - d_{eff})/d_2)}, \text{ and}$$

$$E_{hp} = \sum_{|i-j|>2} \varepsilon_{ij} f(d_{ij}) \text{ where } f(d_{ij}) = \frac{C_2}{1.0 + \exp((d_{ij} - d_0)/d_t)}.$$

The excluded volume term E_{ex} is a soft sigmoidal potential (see Figure 5.2) where d_{ij} is the interatomic distance between two C_α atoms or between two sidechain center of mass atoms C_s, $d_w = 0.1\text{Å}$ which determines the rate of decrease of E_{ex}, $d_{eff} = 3.6\text{Å}$ for C_α atoms and 3.2Å for the sidechain centroids which determine the midpoint of the function (i.e. where the function equals $1/2$ of its maximum value). The constant multiplier C_1 was set to 5.0 which determines the hardness of the sphere in the excluded volume interaction. Similarly, the hydrophobic interaction energy term E_{hp} is a short ranged soft sigmoidal potential (see Figure 5.3) where d_{ij} represents the interatomic distance between two sidechain centroids C_s, $d_0 = 6.5\text{Å}$ and $d_t = 2.5\text{Å}$ which represent the rate of decrease and the midpoint of E_{hp}, respectively. The hydrophobic interaction coefficient $\varepsilon_{ij} = -1.0$ when both residues i and j are hydrophobic, and is set to 0 otherwise. The constant multiplier $C_2 = 1.0$ determines the interaction value and is the equivalent of $1/5$ of one excluded volume violation. The model is not very sensitive to the pair of constants C_1 and C_2 provided that C_1 is considerably larger than C_2.

TABLE 5.1
Residue Parameters for Center of Mass Sidechain Virtual Atoms

Residue Name	Sidechain Bond Length (Angstroms)	Bond Angle (degrees)	Torsion Angle (degrees)	H-P Designation
ALA	1.531	109.625	238.776	H
ARG	4.180	110.156	219.279	P
ASN	2.485	111.156	222.437	P
ASP	2.482	111.160	223.911	P
CYS	2.065	106.938	227.905	H
GLN	3.130	108.423	219.363	P
GLU	3.106	108.577	222.055	P
GLY	0.000	0.000	0.000	P
HIS	3.176	105.977	223.334	P
ILE	2.324	109.945	227.774	H
LEU	2.590	112.273	219.236	H
LYS	3.474	112.711	218.817	P
MET	2.976	113.370	218.790	H
PHE	3.399	112.055	222.650	H
PRO	1.868	64.159	241.896	P
SER	1.897	108.155	237.853	P
THR	2.107	109.617	231.888	P
TRP	3.907	112.930	226.091	H
TYR	3.794	109.695	222.119	H
VAL	1.968	111.792	232.308	H

The final term in the potential energy function, $E_{\varphi\psi}$, is the torsional penalty term allowing only "realistic" (φ, ψ) pairs in each conformation. That is, since φ and ψ refer to rotations of two rigid peptide units around the same C_α atom (see Figure 5.1), most combinations produce steric collisions either between atoms in different peptide groups or between a peptide unit and the side chain attached to C_α (except for glycine). Hence, only certain specific combinations of ε, ψ pairs are actually observed in practice, and are often conveyed via the Ramachandran plot, such as the one for threonine (THR) in Figure 5.4, and the $\varphi - \psi$ search space is therefore very much restricted. The energy term $E_{\varphi\psi}$ accounts for this.

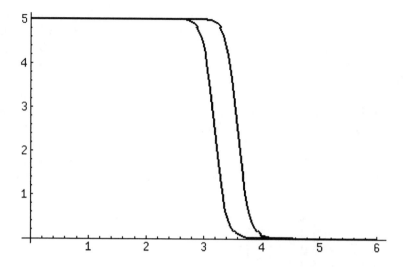

FIG. 5.2. *Excluded Volume Interaction Potential Function Term.*

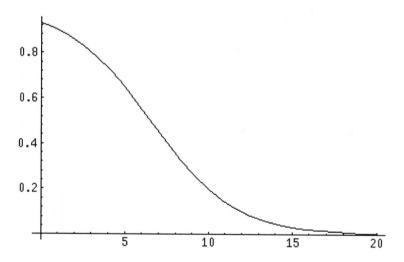

FIG. 5.3. *Hydrophobic Interaction Potential Function Term.*

To compare the simplified and more detailed models at this point, it should be clear that the simplified "string of beads" model treated both φ and ψ together as a single "virtual dihedral angle" (denoted by ξ), thereby reducing the number of independently varying parameters from $2n - 2$ to only $n - 1$ (compare Figure 5.1 with Figure 2.1). In the simplified model, each of the backbone components $NH-C_\alpha i$ $H-C'O$ and the associated sidechain molecules C_α C_s were replaced by a *single* backbone "virtual atom" denoted in Figure 2.1 by C (note that the backbone bond angles θ

and backbone bond lengths l were fixed at their ideal values, even though they *do not* represent the same quantities as shown in Figure 5.1). In addition, because the single dihedral angle ξ effectively replaced the (φ, ψ) pair for each residue, there were no corresponding peptide planes and no sidechain molecules in the simplified model. However, in order to retain the H-P effects that are encoded within the backbone sequence, each "virtual atom" C in that model was categorized as either H or P, thus incorporating the hydrophobic nature of what would have been the associated sidechain (if it were represented). Hence, the CGU algorithm can be applied unchanged to this new more realistic model using the differentiable potential energy function $E_{total}(\phi)$ from equation (2.1) (with the new definitions for E_{ex}, E_{hp}, and $E_{\varphi\psi}$), where $\phi \in \mathbf{R}^\tau$ with $\tau = 2n - 2$ in place of $n - 1$.

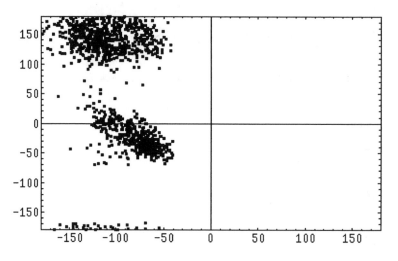

FIG. 5.4. *Ramachandran Plot for Threonine (THR).*

Computational testing of the CGU algorithm using this new detailed polypeptide model with the Sun/Thomas/Dill potential energy function on actual protein sequences is presented in the separate paper by Dill, Phillips, and Rosen [2].

6. Conclusions. Preliminary computational testing of the CGU algorithm applied to a simplified polypeptide model has demonstrated that the method is practical for both the homopolymer and heteropolymer models and for sequences with as many as 48 monomers. Furthermore, since the CGU algorithm is a global optimization method which is not model specific, it can be applied unchanged to the more detailed polypeptide model, or to any other protein model which depends on finding the global minimum of a differentiable potential energy function.

7. Acknowledgments. The authors wish to acknowledge Professor David Ferguson and his colleagues in the Department of Medicinal Chemistry at the University of Minnesota for their valuable contributions on the simple polypeptide model.

REFERENCES

[1] K.A. Dill, *Dominant Forces in Protein Folding*, Biochemistry **29** (1990), 7133–7155.

[2] K.A. Dill, A.T. Phillips, and J.B. Rosen, *Molecular Structure Prediction by Global Optimization*, Journal of Global Optimization, in press (1996).

[3] D.A. Hinds and M. Levitt, *Exploring Conformational Space with a Simple Lattice Model for Protein Structure*, Journal of Molecular Biology **243** (1994), 668–682.

[4] A.L. Lehninger, *Biochemistry: The Molecular Basis of Cell Structure and Function*, Worth Publishers, New York, 1970.

[5] M. Levitt and A. Warshel, *Computer Simulation of Protein Folding*, Nature 253 (1975), 694–698.

[6] C.D. Maranas, I.P. Androulakis, and C.A. Floudas, *A Deterministic Global Optimization Approach for the Protein Folding Problem*, Dimacs Series in Discrete Mathematics and Theoretical Computer Science, (1995), 133–150.

[7] S. Miyazawa and R.L. Jernigan, *A New Substitution Matrix for Protein Sequence Searches Based on Contact Frequencies in Protein Structures*, Protein Engineering **6** (1993), 267–278.

[8] A. Monge, R.A. Friesner, and B. Honig, *An Algorithm to Generate Low-Resolution Protein Tertiary Structures from Knowledge of Secondary Structure*, Proceedings of the National Academy of Sciences USA **91** (1994), 5027–5029.

[9] A.T. Phillips, J.B. Rosen, and V.H. Walke, *Molecular Structure Determination by Convex Global Underestimation of Local Energy Minima*, Dimacs Series in Discrete Mathematics and Theoretical Computer Science **23** (1995), 181–198.

[10] J. Skolnick and A. Kolinski, *Simulations of the Folding of a Globular Protein*, Science **250** (1990), 1121–1125.

[11] R. Srinivasan and G.D. Rose, *LINUS: A Hierarchic Procedure to Predict the Fold of a Protein*, PROTEINS: Structure, Function, and Genetics **22** (1995), 81–99.

[12] S. Sun, *Reduced representation model of protein structure prediction: statistical potential and genetic algorithms*, Protein Science **2** (1993), 762–785.

[13] S. Sun, P.D. Thomas, and K.A. Dill, *A Simple Protein Folding Algorithm using a Binary Code and Secondary Structure Constraints*, Protein Engineering **8** (1995), 769–778.

[14] K. Yue, K.M. Fiebig, P.D. Thomas, H.S. Chan, E.I. Shakhnovich, and K.A. Dill, *A Test of Lattice Protein Folding Algorithms*, Proceedings of the National Academy of Sciences USA **92** (1995), 325–329.

POTENTIAL TRANSFORMATION METHOD FOR GLOBAL OPTIMIZATION

ROBERT A. DONNELLY*

Abstract. Several techniques for global optimization treat the objective function f as a force-field potential. In the simplest case, trajectories of the differential equation $m\ddot{\mathbf{x}} = -\nabla f$ sample regions of low potential while retaining the energy to surmount passes which might block the way to regions of even lower local minima. A *potential transformation* is an increasing function $V: \mathbf{R} \to \mathbf{R}$. It determines a new potential $g = V(f)$, with the same minimizers as f, and new trajectories satisfying $m\ddot{\mathbf{x}} = -\nabla g = -\frac{dV}{df}\nabla f$. We discuss a class of potential transformations that greatly increase the attractiveness of low local minima. As a special case, this provides a new approach to an equation proposed by Griewank for global optimization. Practical methods for implementing these ideas are discussed, and the method is applied to three test problems.

1. Introduction. On the following pages we consider a method for global optimization which is based on application of a local transformation to the objective function f whose minima are sought. The impetus for this transform comes from earlier work of Andreas Griewank [9], who presented a second-order ordinary differential equation which could be used for global optimization:

$$(1.1) \qquad \ddot{\mathbf{x}} = \left[-\epsilon\nabla f + (1+\epsilon)\frac{\dot{\mathbf{x}}\cdot\nabla f}{\|\dot{\mathbf{x}}\|^2}\dot{\mathbf{x}} \right](f-c)$$

It governs the trajectories followed by a unit-mass particle in search of local minimizers of f. Two heuristics of this equation are especially significant: First, the particle speed can be shown to be $f - c > 0$ where c represents a *target level* of optimization; Second, trajectories are continually accelerated in the direction of the negative gradient under the control of a "sensitivity parameter" ϵ. The geometry of Griewank's equation is shown in Fig. 1.1 for two different choices of gradient sensitivity. One sees that increasing ϵ has the effect of changing the direction of acceleration toward that of steepest descent.

Initially we judged the use of differential equations for optimization to be inefficient, inasmuch as we felt that accurately tracking the trajectories of a differential equation imposes an unreasonably high computational overhead in problems involving more than a few independent variables. Thus we incorporated the heuristics and geometry of Griewank's equation into a discrete algorithm known as SNIFR. We briefly review the algorithm and some of its applications in the following Section. Experience with SNIFR indicated that it might be useful to return to a continuous setting, especially for problems involving integrable constraints, for these can often be nicely treated using Lagrange multipliers [21]. This led us to reconsider

* Chemistry Department, Auburn University, Auburn, AL 36830.

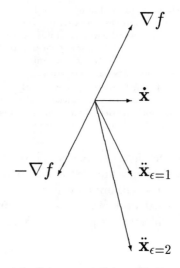

FIG. 1.1. *Geometry of Griewank's Equation*

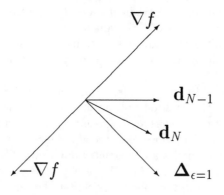

FIG. 1.2. *Geometry of SNIFR*

Griewank's equations, with the result that we can now derive them from Newton's equation of motion for a particle moving under the influence of a potential. In fact we can show [20] that Griewank's equations are but one special case of the ordinary differential equation which forms the basis for what we now call the *potential transform* or **PT** method. We sketch our derivation of this, following the derivation with an evaluation of PT on a few test problems.

We begin with a problem which comes from molecular chemistry, pedagogically useful as it includes the full range of interactions found in typical molecules. Relegating a detailed description of the objective function to an Appendix we consider applications of SNIFR, PT, and Molecular Dynamics to determination of the global energy minimum of the cyclohexane molecule. We then briefly consider two problems taken from a recent publication in which we compared the efficacy of five global optimization algorithms [20]. The first parallels a problem proposed by Griewank, but is much harder than this one. The second again comes from chemistry, in which one attempts to find the global minimum of a system of N atoms forming a microcluster under the influence of the Lennard-Jones potential. We finish with a short summary of results. Special thanks go to Jack W. Rogers, Jr. for his original derivation of the PT equation, and to Andreas Griewank for thoughtful discussions.

2. SNIFR. The SNIFR algorithm is a discrete dynamical system modeling Griewank's heuristics (see Fig. 1.2). In it we replace velocity and acceleration vectors $\{\dot{\mathbf{x}}, \ddot{\mathbf{x}}\}$ by a new set $\{\mathbf{d}, \boldsymbol{\Delta}\}$ and apply Griewank's heuristics. Assume the search direction on the $(N-1)^{\text{st}}$ step was the *unit* vector $\mathbf{d}_{N-1}$ and take the rotation $\boldsymbol{\Delta}$ as

$$\boldsymbol{\Delta} = -\epsilon\nabla f + (1+\epsilon)(\mathbf{d}_{N-1} \cdot \nabla f)\mathbf{d}_{N-1}$$

We turn $\mathbf{d}_{N-1}$ toward the steepest descent direction using parameter ϵ by defining the direction of step N to be a new unit vector

$$\mathbf{d}_N = \frac{\boldsymbol{\Delta} + \mathbf{d}_{N-1}}{||\boldsymbol{\Delta} + \mathbf{d}_{N-1}||}$$

The coordinates are updated by taking a step along $\mathbf{d}_N$ of length β where

$$\beta = max\left[\beta_{min}, min\left[\mu(f - c), \beta_{max}\right]\right],$$

i.e. $\mathbf{x}_N = \mathbf{x}_{N-1} + \beta\mathbf{d}_N$. Incorporating Griewank's heuristics into a discrete dynamical system obviates the need to integrate Eq. 1.1. The discrete system is governed by parameters $c, \epsilon, \mu, \beta_{min}$ and β_{max}, where c and ϵ have their previous meanings. μ is a proportionality constant which may be used to scale values of $f - c$, and in practice we limit the stepsize to the interval $\beta \in \{\beta_{min}, \beta_{max}\}$ on account of the discrete nature of the algorithm. Otherwise – depending on the roughness of the surrounding terrain– one

TABLE 2.1
Simulated Annealing(SA) vs SNIFR(SN)

Function	SA		SNI		SNII	
	N*	%**	N	%	N	%
F1	3910	54	3695	90	1040	54
F2	3421	64	2655	96	1092	64
F3	3078	81	3070	95	1589	81
F4	1224	100	534	99	534	99
F5	1914	62	1760	99	364	54
F6	557	100	205	100	205	100
F7	1186	99	664	100	664	100

* Number of function/gradient evaluations
** Success rate

might take an unacceptably large step when $f - c$ is large. Alternately, creeping slowly along, even in areas where $f - c$ is small is inefficient; we prefer to keep a minimum stepsize, and typically take $\beta_{min} = .05\beta_{max}$. Since the global minimum of the objective function is often unknown, one might think that setting the target is difficult. We have not found this to be the case: we routinely set c arbitrarily at first. Then we continually frustrate the system by *resetting the target* should a trajectory approach it too closely.

In spite of its simplicity SNIFR has turned out to be a remarkably successful global optimization algorithm. It has been used in studies of reentry-vehicle trajectories [17], in celestial mechanics [24], and in refinement of protein structures utilizing interparticle-distance constraints derived from Nuclear Magnetic Resonance spectroscopy [10]. Table 2.1 summarizes the results of one study by Butler and Slaminka [5], in which they assessed the performance of SNIFR versus Simulated Annealing [14] on a suite of seven test problems proposed in Dixon [7]. The first column represents number of function evaluations and the success rate achieved by Vanderbuilt and Louie [29]. The second and third columns refer to function/gradient evaluations and success rates achieved by Butler and Slaminka with each of two procedures. In the first, labeled SI, they applied the SNIFR algorithm, adjusting system parameters to achieve a minimum 90% success rate. Series SII was obtained by making minor adjustments in the SI parameters in order to attain the same success rate reported by Vanderbuilt and Louie. SNIFR compares favorably to Simulated Annealing, returning either better success rates, or comparable success rates with at worst 56% of the function evaluations required in Vanderbuilt and Louie. Details of the implementation of SNIFR are given in Ref. [5].

3. The PT equations.

3.1. Newton's equation. Several optimization strategies are based on Newtonian dynamics [25,15]. One takes the view that the function $f(\mathbf{x})$ whose minima are sought generates a force $\mathcal{F} = -\nabla f$ acting on a particle of mass m; trajectories followed by this particle satisfy Newton's equation of motion

$$(3.1) \qquad\qquad m\ddot{\mathbf{x}} = -\nabla f$$

subject to conservation of total energy

$$(3.2) \qquad\qquad E = T + f$$

where $T = (1/2)mv^2$, the kinetic energy, is proportional to the temperature of the system. It is convenient to remove the explicit mass dependence in Newton's equation. Replace each x there by $x\sqrt{m}$. Then $dx \rightarrow \sqrt{m}dx$ and Newton's equation becomes

$$(3.3) \qquad\qquad \ddot{\mathbf{x}} = -\nabla f.$$

Eq. 3.3 is useful for optimization because the dynamical rules prescribe acceleration in the direction of steepest descent leading to low values of the objective function. Higher values are accessible to the system by virtue of conservation of energy, so that hills separating adjacent valleys can often be traversed in the search for low values of f. *Annealing* the system [6] reduces the kinetic energy via a time-dependent *annealing schedule* $T = s(t)$ which reduces the particle speed (and hence the total energy) as a function of time. Thus the likelihood that the system is trapped by *some* local minimum increases with simulation time. Though in certain cases it can be shown [2] that continuous simulated annealing can produce global minimizers the required simulations are perforce long and expensive.

Of course, one problem with Newtonian heuristics *vis a vis* optimization is the fact that the system *races* through minima while s-l-o-w-i-n-g in regions where the potential is high. In fact, one might desire just the opposite behavior: equations which prescribe high speeds in regions where the potential is high, and low values in regions where it is low.

3.2. Derivation of PT. The Newtonian speed is constrained by energy conservation, $v = \sqrt{2(E-f)}$. Incorporation of Griewank's first heuristic involves replacing the Newtonian speed by the new speed $v = (f - c)$. Thus it simplifies matters to recast Newton's equation in arclength parameterization, since this is a unit-speed parameterization:

$$\mathbf{x}' = \frac{d\mathbf{x}}{ds} \; ; \|\mathbf{x}'\| = 1 \; ; \mathbf{x}' \cdot \mathbf{x}'' = 0.$$

The derivatives $\dot{\mathbf{x}}$ and $\mathbf{x}'$ are related by $\dot{\mathbf{x}} = v\mathbf{x}'$ where v is the Newtonian speed: in terms of $\mathbf{x}'$ and $\mathbf{x}''$ one may write Eq. 3.3 as

$$(3.4) \qquad\qquad \ddot{\mathbf{x}} = \frac{dT_v}{ds}\mathbf{x}' + 2T_v\mathbf{x}''$$

where T_v is the kinetic energy. The orthogonality of $\mathbf{x}'$ and $\mathbf{x}''$ suggests decomposition. Let $\mathbf{P_{x'}\ddot{x}}$ be the projection of $\ddot{\mathbf{x}}$ on $\mathbf{x}'$

$$\mathbf{P_{x'}\ddot{x}} = \frac{(\ddot{\mathbf{x}} \cdot \mathbf{x}')}{\|\mathbf{x}'\|^2}\mathbf{x}'$$

and $\mathbf{Q_{x'}\ddot{x}}$ be its complement. Then

$$\ddot{\mathbf{x}} \equiv \mathbf{P_{x'}\ddot{x}} + \mathbf{Q_{x'}\ddot{x}}$$

from which we obtain Newton's equation in arclength parameterization:

$$(3.5) \qquad\qquad \mathbf{x}'' = \frac{1}{2T_v}\mathbf{Q_{x'}}\nabla f$$

The norm of $\mathbf{x}''$ is the radius of curvature κ of the trajectory.

We now suppose that we replace the original potential f by a new one $V(f)$ which we shall in practice choose to make values of f near c more attractive than the corresponding regions in the untransformed system. Further, let V be a strictly increasing function of f, so that minima and maxima of the new potential occur at precisely the *same positions* as those of the original potential. The effect of all this is to replace the original system by one with a new energy $E = T_v + V(f)$. Newton's equation of motion for the new system is

$$(3.6) \qquad\qquad \ddot{\mathbf{x}} = -\nabla V(f) = -\frac{dV}{df}\nabla f$$

subject to conservation of energy. The unit speed parameterization of this equation is

$$(3.7) \qquad\qquad \mathbf{x}'' = -\frac{dV}{df}\frac{1}{2T_v}\mathbf{Q_{x'}}\nabla f.$$

The radius of curvature is

$$(3.8) \qquad\qquad \kappa = \frac{\frac{dV}{df}}{2T_v}\|\mathbf{Q_{x'}}\nabla f\|.$$

We shall return to this important equation shortly.

Now we derive the Newton's equation for an arbitrary speed parameterization by starting with the unit-speed one. Let us transform variables $\{t, v = \frac{ds}{dt}, \dot{\mathbf{x}} = \mathbf{x}'v\} \rightarrow \{\tau, \sigma = \frac{ds}{d\tau}, \overset{*}{\mathbf{x}} = \mathbf{x}'\sigma\}$ so that Eq. 3.4 becomes

$$(3.9) \qquad\qquad \overset{**}{\mathbf{x}} = \frac{dT_\sigma}{ds}\mathbf{x}' + 2T_\sigma\mathbf{x}''$$

in which the new kinetic energy is $T_\sigma = \sigma^2/2$, and the star signifies differentiation with respect to τ. Provided we require the new speed σ to be a function of f, application of the chain rule to the first term in Eq. 3.9 gives

$$\frac{dT_\sigma}{ds}\mathbf{x}' = \frac{dT_\sigma}{df}(\nabla f \cdot \frac{d\mathbf{x}}{ds})\mathbf{x}' = \frac{dT_\sigma}{df}\mathbf{P_{x'}}\nabla f$$

Retaining the curvature from Eq. 3.7 we finally obtain the equation of motion associated with the potential transformation:

$$(3.10) \quad \overset{**}{\mathbf{x}} = -\frac{dV}{df}\frac{T_\sigma}{T_v}\nabla f + \left[\frac{dT_\sigma}{df} + \frac{dV}{df}\frac{T_\sigma}{T_v}\right]\mathbf{P_{x'}}\nabla f \quad \textbf{PT equation}$$

This is essentially Newton's equation of motion for a unit mass particle moving under the influence of a potential $V(f)$ with conserved total energy and speed $\sigma(f)$.

An entire family of optimization methods an be generated by different choices for the potential transform and speed. Of course, with $\{V(f) = f, \sigma = v, T_\sigma = T_v = E_0 - V\}$ one recovers, Eq. 3.3. Griewank chose a potential of the form

$$(3.11) \qquad\qquad V(f) = \frac{-1}{(f-c)^{2\epsilon}}.$$

He prescribed a speed $\sigma = f - c$ but did not consider energy conservation in his equations, since he did not perceive them as derivable from Eq. 3.3. With these substitutions, and taking $E_0 = 0$, Eq. 3.10 becomes

$$\overset{**}{\mathbf{x}} = \left[-\epsilon\nabla f + (1+\epsilon)\mathbf{P_{\overset{*}{x}}}\nabla f\right](f - c)$$

which was stated in Ref. [9].

One might argue that, instead of integrating Eq. 3.10, one could simply integrate Eq. 3.6 directly, for any particular choice of V. However, direct integration could be quite costly when V is varying rapidly, or when evaluation of the objective function is expensive. Rather than spend time in generating extremely accurate trajectories using special integration methods, we prefer to use inexpensive low-order methods. We argued in [20] that the natural choice for the speed is to make it low when the radius of curvature in Eq. 3.8 is high. We showed that when the potential is the particular one chosen by Griewank(Eq. 3.11) the speed is given by

$$(3.12) \qquad\qquad \sigma = (f-c)(1 - \frac{E_0}{V}).$$

The energy(which is conserved) is chosen so that the trajectory can sample values of the objective function between c and f_{max} by the choice $E_0 = V(f_{max})$. This makes the speed approach zero as $f \to c$ or $f \to f_{max}$. With these substitutions Eq. 3.10 becomes the working equation for the numerical examples considered in the next Section

$$(3.13) \qquad \overset{**}{\mathbf{x}} = \left[-\epsilon\nabla f + (1+\epsilon - (1+2\epsilon)\frac{E_0}{V})\mathbf{P_{\overset{*}{x}}}\nabla f\right]\sigma.$$

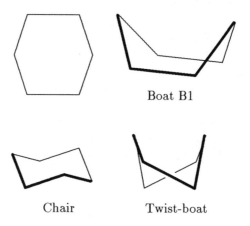

Boat B1

Chair   Twist-boat

FIG. 4.1. *Cyclohexane structures(schematic)*

4. Evaluation of PT.

Problem 1: Molecular conformations. We begin our evaluation of the PT method with a pedagogical problem illustrating one application of minimization methods to chemistry, namely search for the molecular conformation which is also the global energy minimum. The term "molecular conformation" refers to different arrangements of atoms produced by rotation about chemical bonds: a change in conformation does not involve breaking chemical bonds. Mathematically speaking, "conformations" are stationary points on a potential energy surface at which the potential energy is a (local) minimum. A "force field" is a potential energy function which has been constructed to reproduce experimentally-observed conformations of (usually small) molecules. The potential energy depends on the Cartesian coordinates of each of the atoms in a manner which is described more completely in the Appendix. Our strategy is to begin with a more-or-less random approximation to a high-energy molecular structure and to search for the global minimum by application of Newtonian Dynamics (also called molecular dynamics or MD), transformed dynamics (PT), and SNIFR.

The cyclohexane molecule is formed by a hexagonal ring of six carbon atoms, each bonded to two neighboring carbons, and each bearing two hydrogen atoms. Rather than treat all the atoms as individuals it is customary to represent each CH_2 as a single unit ("united atom"): each vertex in Fig. 4.1 is occupied by a single CH_2 group. The molecule can exist in either of two conformations, the *chair* or the *twist-boat* With an energy

ϵ of 27 *units* the chair is the global minimum; the twist-boat has an energy of roughly 69 *units* using parameters taken from GROMOS [28]. The chair and twist-boat forms of cyclohexane can be interconverted through its "transition state," which resembles a boat(BI in Fig. 4.1). A "transition state" is a saddle point on a potential surface. The actual boat has an energy that is higher than that of either chair or twist-boat forms.

Starting structures for each of two sampling trajectories trajectory were constructed by taking a regular planar hexagon and lifting two opposing vertices (Fig. 4.1). Series SI used structure *BI* directly; its potential energy is $\epsilon = 354$ *units*. Since *BI* was rather strained, we applied steepest descents to remove some of the strain energy, producing structure *BII* with an energy of 100 *units*. *BII* was used to start a second search trajectory SII. We confirmed that both steepest descent and conjugate gradients converged to the higher-energy twist-boat when begun at either *BI* or *BII*. Thus we may terminate a search trajectory when any method achieves an energy significantly lower than that of the twist-boat, since a strictly descending local optimization method would find only the global chair minimum. when started with a structure whose energy was lower than that of the twist-boat.

The MD,SN, and PT algorithms were applied using Euler's method to integrate the MD or PT differential equations for 1000 steps with a time step of $\delta t = 1 \times 10^{-4}$ *units*. Alternately we used 1000 steps of the SNIFR algorithm with a maximum stepsize $\beta_{max} = 1 \times 10^{-3}$. PT and SNIFR used a gradient sensitivity $\epsilon = 1$ and a fixed target $c = 50$ *units*. Define $f_{max} \equiv f(\mathbf{x}_0) + 1$ where $\mathbf{x}_0$ is the vector of starting coordinates for the molecular geometry. Then the total energy was taken as $E_0 = V(f_{max})$ in the PT calculations and as $E_0 = f_{max}$ in the MD calculations. Thus either system will have zero kinetic energy when the potential reaches f_{max}, which makes structures with potential energies greater than this value inaccessible. All algorithms were started in the steepest descent direction, with initial speeds $\sigma = (f_0 - c)(1 - V(f_{max})/V(f_0))$ for PT or $v = \sqrt{(2(E_0 - f_0))}$ for the MD calculations; SNIFR was begun with speed $(f_0 - c)$.

Results for series SI and SII are displayed graphically in Figs. 4.2 and 4.3, where we plot potential energy as a function of step number for the respective methods and starting points. Only the first 100 steps of SI are shown since even by step 1000 only PT succeeded in getting below the potential of the twist-boat; steps 300 through 600 are shown for SII, where again PT was the only method to find energies lower that that of the twist-boat. Though the high total energy($E_0 = 355$ *units*) in series SI allowed MD to sample structures with a wide range of potential energies this trajectory could not find its way into the basin of attraction of the chair. The lower total energy ($E_0 = 101$ *units*) in series SII produces smaller energy fluctuations than were observed in the first trajectory for both MD and PT, so (Fig. 4.3) PT took a correspondingly longer time to find the basin of attraction of the chair. In both series SNIFR was rather disappointing: it settled down rather rapidly to an orbit of the twist-boat

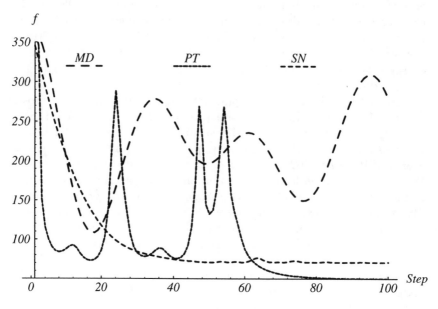

FIG. 4.2. *Part of Trajectory Number SI*

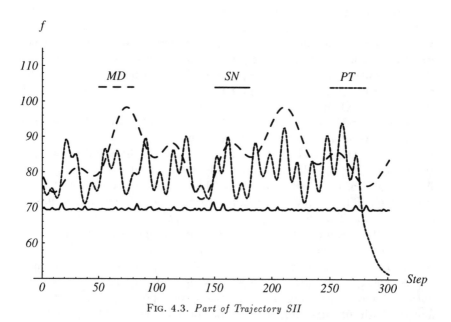

FIG. 4.3. *Part of Trajectory SII*

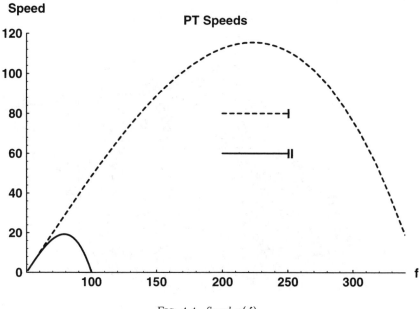

FIG. 4.4. *Speed* $\sigma(f)$

form, where it remained trapped for the duration of its run. In fact it is just such behavior of SNIFR which caused us to reexamine its theoretical underpinning, leading to our derivation of the PT equations! In both series PT made some large uphill moves and found the basis of attraction of the chair with relative ease. We attribute this in part to the distinctly non-Newtonian character of the speed profile available to PT (Fig. 4.4). PT is able to speed up as it climbs moderately-sized hills, thus quickly surmounting them. Should the energy become too high PT again slows, allowing it to be diverted into lower energy regions. Should it near the target PT settles down and thoroughly covers areas where $f - c$ is small.

Problem 2: The p-function. Recently we compared the efficacy of five global optimization methods (PT, Multi-Start(MS) [26], Newtonain Dynamics(MD) [25,4], Simulated Annealing(SA) [14], and SNIFR(SN) [8], [19]) on two different test problems. Here we summarize some of the results we obtained using MD, SN, and PT. Details of the procedures followed, as well as comparisons to Simulated Annealing and Multi-Start can be found in reference [20].

The first function we considered uses the *p*-norm of a vector,

$$\|\mathbf{x}\|_p = \left(\sum_{i=1}^{n} |x_i|^p \right)^{1/p},$$

Untransformed

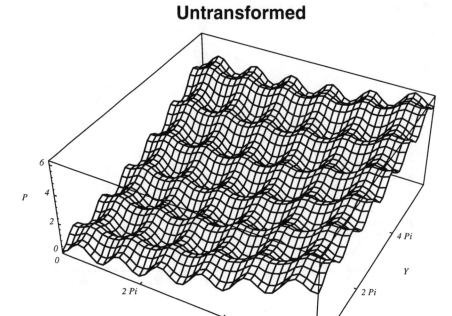

FIG. 4.5. *The p-function*

where $p \geq 1$. Define the C^1-function $f : \mathbf{R}^n \to \mathbf{R}$ by

$$(4.1) \qquad f(\mathbf{x}) = \frac{\|\mathbf{x}\|_p^2}{\lambda} + \|\sin \mathbf{x}\|_p^2 \,,$$

where the sin of a vector is taken component-wise, $\lambda = 90$ and $-\frac{13\pi}{2} \leq x_i \leq \frac{13\pi}{2}$. This we call the "p-function."

The values of f along an axis and along a diagonal are are *independent of dimension*, which indicates that the shape of the function in high dimensions — and its difficulty as a global optimization test function — should be similar to that in low dimensions. A hidden-surface plot is shown for two independent variables in Fig. 4.5. It has a single global minimum at the coordinate origin and a number of sub-optimal local minima which grows exponentially with dimension. Local minimizers occur on or near the boundary of n-cubes centered at the origin There are thirteen such minimizers in the range $-\frac{13\pi}{2} \leq x \leq \frac{13\pi}{2}$, and it can be shown that the minimizers are grouped into seven bins, $b_0, \ldots, b_6$, with only the origin in b_0. The bin number assigned to a given local minimizer is a simple measure of how close the minimum is to the global one.

MD, PT, and SN trajectories were run for the p-function in several

TABLE 4.1
MD, PT, SN comparison

BIN	0	1	2	3	4	5	6	Dim
MD		13	22	13	1			
PT	31	15	3	1				
SN	49	1						5
MD			3	20	23	4		
PT	38	10		2				
SN	50							10
MD				1	19	26	4	
PT	18	31			1			
SN	39	11						20
MD				2	4	34	10	
PT	1	26	23		1			
SN	5	45						30
MD					3	28	19	
PT		50						
SN		50						50

dimensions following procedures established in the previous section. PT used a gradient sensitivity $\epsilon = 1$. and target level $c = -.2$. A time-step of $\delta t = .4$ was used by both PT and MD; f_{max} was set at 30 for both. SNIFR used a gradient sensitivity $\epsilon = 1$, maximum steplength $\beta_{max} = 1$, and a fixed target, $c = -.2$. All trajectories consisted of 32,000 steps from each of 50 random starting points selected from the range $-\frac{13\pi}{2} \leq x \leq \frac{13\pi}{2}$.

At first, 32,000 steps (function/gradient calculations) may seem excessive in these low-dimensional problems (d = 5,10,20,30,50 dimensions). However, recall Figure 4.5. Given such hilly terrain it should be difficult for any method to find the global minimum. In 10 dimensions, for example, there are roughly 10^{11} local minima of the p function. The cumulative probability of finding a minimum in bins $b_0, b_1, b_2, \cdots, b_j$ in d dimensions after s steps is

$$P_j(s,d) = 1 - \left(1 - \left(\frac{(2j+1)}{13}\right)^d\right)^s \approx s(2j + 1/13)^d$$

or approximately $P_0 = 2 \times 10^{-7}, P_1 = .0137, P_2 = .8964$ after 32,000 steps.

In light of these considerations our results, collected in Table 4.1, show that PT performs remarkably well. In 10 dimensions one observes that PT finds the global minimum fully 76 % of the time, and is otherwise in bins 0, 1 96 % of the time. In 50 dimensions the p function has roughly 5×10^{55} local minima, and the probability of finding a minimum in bins 0, 1, 2 using a single run of 32,000 steps is roughly $P_2 = 6 \times 10^{-17}$. We observe that PT finds minima in bins 0, 1, 2 100 % of the time. By contrast,

Transformed

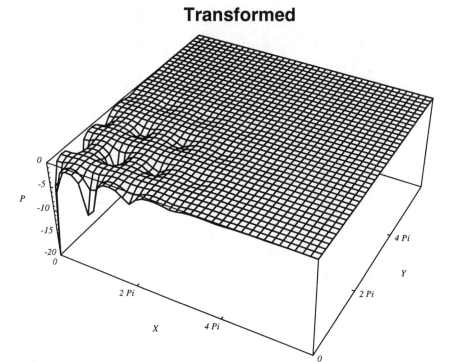

FIG. 4.6. *Transformed p-function*

ordinary (i.e. *untransformed*) Newtonian dynamics performs poorly on the p-function. In fact (see the original source) MD performs even worse than Multi-Start did on this problem. From this we conclude that the effects of the potential transform are indeed potent. Table 4.1 shows that SNIFR behaves much better on this problem than it did on Problem 1. It almost always finds the global minimum for problems with dimensions ranging up to about $d = 20$. Beyond this dimension its performance deteriorates, though by dimension 50 one still finds that all minima are in bins 0, 1. SNIFR was the most effective algorithm we tried on the p-function.

The failure of PT to find the global minimum with greater probability stems from the fact that it, too, can be trapped by sub-optimal minima, though with much lower frequency than one observes in MD. One reason for this trapping can be perhaps suggested by Figure 4.6 in which we display the transformed function in two dimensions for the current choice of system parameters ($\epsilon = 1, c = -.2, f_{max} = 30$). Minima further out from the origin than those in bins 0, 1, 2 have virtually disappeared under the effects of the potential transform, while those nearer the origin remain, and are rather deep. Thus should the particle come near the origin, our data indicates that it remains trapped in one of the sub-optimal minima for the remainder of

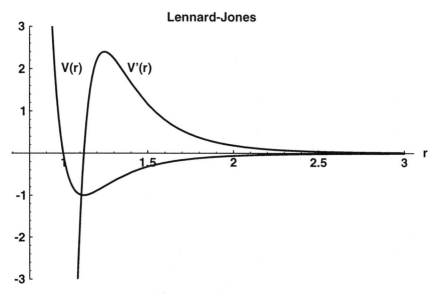

FIG. 4.7. *Lennard-Jones Potential*

the trajectory.

Problem 3: Microclusters. The final problem we consider is determination of the ground-state of a system of unit-diameter spheres interacting via the Lennard-Jones potential. Let r_i denote the Cartesian coordinates of sphere i, and let $\rho_{i,j}$ be the vector $r_i - r_j$. Then the Lennard-Jones potential [3] is defined by the formula

$$(4.2) \qquad v(\rho_{i,j}) = \|\rho_{i,j}\|^{-12} - \|\rho_{i,j}\|^{-6}$$

a sketch of which is shown in Fig. 4.7 The equilibrium interparticle separation of a pair occurs at $\rho = 2^{\frac{1}{6}}$, repulsive forces dominate at shorter distances, while attractive forces dominate out to about $\rho = 3$. A system of N such spheres has the potential energy

$$(4.3) \qquad V = \sum_{i=1}^{N-1} \sum_{j=i+1}^{N} v(\rho_{i,j}).$$

Such a potential is expected to provide a tractable model of the many-body potential experienced by a cluster of real atoms.

Early work of Hoare and Pal([12,13] established certain regularities in the structures of atomic clusters modeled by the Lennard-Jones potential, as well as the existence of large numbers of local minimizers. Hoare and McInnes [11] exhaustively studied the problem for cluster sizes ranging from 6 to 13 spheres, reporting the following number of potential-energy minima: 2;4;8;18;57;145;366;988. Thus one sees that the number of cluster configurations(local minima) rises much faster than linearly, for the number of minima, g, is fit by

$$(4.4) \qquad g(N) = \exp(-2.5176 + .3572N + .028N^2)$$

A small cluster with 15 atoms should then have something on the order of 10,750 local minima; a cluster containing 25 atoms is expected to have somewhere in the neighborhood of 10^{10} local minima! This explosive growth of cluster configurations makes space covering –or other exhaustive techniques– quite impossible for all but the smallest clusters.

Others soon followed, occasionally reporting better local minimizers for even small clusters. Northby [16] developed a systematic search procedure which he then successfully applied to clusters containing up to 150 atoms. Tsai and Jordan [27] found 1328 local minima for the Ar_{13} cluster, where Hoare and McInnes had reported finding 988(see also Kunz and Berry in [18]). Very large clusters have been studied by Xue [30,31] using simulated annealing, while Coleman, Shalloway, and Wu approached the problem using the build-up methdology [23]. Other approaches were reviewed recently by Pardalos, Shalloway, and Xue [22].

In view of these studies we were especially interested to know how the PT method performed as a function of the number of steps taken, and as a function of the gradient sensitivity. Thus we studied clusters containing up to 25 atoms while varying the gradient sensitivity. Rather than use a methodology tailored to the special structure of the cluster problem as had been done by previous authors we instead chose initial coordinates for each atom randomly on a cube of edgelength three units (remember that the potential is negligible beyond 3 units). Starting coordinates which would have produced overlapping spheres were discarded, due to the strongly repulsive behavior of the forces near the origin refer to Fig .4.5. The minimum found in each run was subsequently polished using conjugate gradients. Results are summarized in Table 4.2, where entries labeled $N = 0$ refer to a local minimum obtained by direct application of conjugate gradients to the starting point. The lowest minima reported by Northby [16] are labeled by $N = \infty$ in the Table. These seem to have stood the test of time and may (at least as of 1995) be regarded as the global minimizers the indicated number of spheres.

One sees that with a modest amount of work we acchieved the global minima in all cases we studied. In fact, the number of function/gradient evaluations is quite small considering the complexity (i.e. the number of local minima) of the objective function. The largest system we studied contains 25 particles, for which we obtain the global minimum with only 30,000 function/gradient calls. Results for the 16-atom cluster show clearly that slight changes in the gradient sensitivity can have a strong influence on the minima obtained. Note especially that increasing ϵ does not always produce lower minima: increasing this parameter too much can lead to undesirable "orbiting" of local minima(see [19] for interesting pictures regarding orbit-

TABLE 4.2
Sphere-Packing Results

14	N	ϵ	-Energy
	0		44.880
	∞		47.845
	1000	1.	47.845
16	0		55.907
	∞		56.816
	1,000	.9	53.757
	1,000	1.	52.807
	1,000	1.5	54.941
	2,000	.9	55.195
	2,000	1.	54.909
	2,000	1.5	54.941
	5,000	.9	56.816
	5,000	1.	56.816
	5,000	1.5	56.816
23	0		89.696
	∞		92.844
	5,000	.8	90.265
	5,000	.9	92.844
	5,000	1.	91.198
	5,000	1.1	90.984
24	0		91.194
	∞		97.349
	5,000	.9	96.239
	5,000	1.	96.517
	10,000	.9	97.349
	10,000	1.	96.517
25	∞		102.37
	10,000	.9	96.533
	30,000	.9	101.08
	10,000	1.	99.528
	30,000	1.	102.37
	30,000	1.5	101.08

ing of minima). Reducing ϵ too much has the effect of reducing sensitivity of the dynamics to the local gradient Finally we should point out that these results were obtained by treating the objective function more or less as a "black box," in the sense that no special methods were followed which exploited the special nature of this problem. We would not argue that this is the correct approach to follow in solving a "real" problem. We would argue, however, that such success rates following an uneducated starting point speaks well for the efficacy of the potential transform method.

5. Summary. On these pages we summarized our recent experience with a new global minimizer which is based on what we feel are potent heuristics. Central to this work was Griewank's equation, which was shown to be but one example of a much wider class of ordinary differential equations which can be used for global optimization. We gave an example of a particular local transformation which we then applied to three test problems.

Each of the methods we developed has strengths and weaknesses which should be exploited when approaching a given optimization problem. Unconstrained, smooth problems are probably best treated by SNIFR. Since in a smooth problem the steplength could be chosen more or less arbitrarily, it is likely that SNIFR is more efficient than integration of a differential equation. SNIFR was not set up to handle constrained problems, so these should be treated by PT. SNIFR was particularly successful on the p-function, and has performed well the singular cluster problem. On the other hand, it appears that the strong, bonded forces in cyclohexane are not so well handled by SNIFR, as its performance was surpassed by PT.

The typical way that the chemical community enforces bond-length constraints in molecular dynamics simulations, for example, is to incorporate them via time-dependent Lagrange multipliers. Thus, let σ represent a set of time-dependent constraints and λ be a set of time-dependent Lagrange multipliers introduced to enforce them. Then the forces of constraint introduce an added term into Newton's equation of motion, *viz:*

$$m\ddot{\mathbf{x}} = -\nabla \mathbf{V} - \lambda^{\mathbf{T}} \nabla \sigma.$$

This, and the wide range of forces encountered in molecular problems(see Appendix) motivated a return to a continuous setting, setting the stage for a derivation of the PT equations. A return to the continuous setting has merit in view of the fact that PT outperformed SNIFR on the cyclohexane problem.

The problem of global minimization is known to be unsolvable, and the claim that a candidate minimizer is a global one remains generally undecidable. Our experience to date shows that the potential transform method is a useful addition to other heuristic methods designed for global optimization.

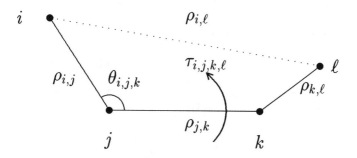

FIG. A.1. *Geometrical parameters*

A. Appendix. Here we describe the molecular energy function used in the test calculation described in the text. A comprehensive set of definitions of the chemical terms we use can be found in Ref. [1]. A comprehensive description of molecular dynamics and molecular potential functions is the book by McCammon and Harvey [15]. and the GROMOS reference manual [28]. As the test molecule is simple we shall gloss over many fine details: consult the references for these.

See Fig. A.1 for a picture of the interactions we describe. Atoms are extensionless points, which possess a mass m and optionally a charge q; a molecule is a collection of atoms, which may be either "bonded" (joined by strong chemical forces represented by solid lines in the Figure) or "nonbonded" (interacting weakly through space, dotted line in the Figure). Pairs of bonded atoms interact via "bond-stretching" forces, which are usually quite strong. Weaker forces exist among triples of atoms in which two atoms are both bonded to a third forming a "valence-angle" $\theta_{i,j,k}$. Also relatively weak is the interaction made by twisting around a bond. Quartets of atoms can form a "dihedral" or "torsion" angle $\tau_{i,j,k,\ell}$, which turns about the bond between atoms j and k. The non-bonded interactions are of the 2-body type. The strongest is the "Coulomb" interaction between atoms bearing a net charge (there are none of these in the cyclohexane example). The Lennard-Jones or "6-12" interaction describes the much weaker forces which act between atoms which are not directly bonded. Though weak, such interactions are quite numerous in the general case, so their inclusion is significant. There may be other, special, terms present in protein forcefields. For now, the interactions described will by sufficient for our purposes.

The energy is a sum of the various terms:

$$\epsilon = \epsilon_{Bond} + \epsilon_{Coulomb} + \epsilon_{Lennard-Jones} + \epsilon_{Angle} + \epsilon_{Dihedral}$$

For particles i, j, k, and ℓ with positions ρ_i, ρ_j, ρ_k, ρ_ℓ, inter-atomic distances $\rho = \|\rho_i - \rho_j\|$, valence angles θ

$$\theta = \cos^{-1}\left[\frac{\rho_{i,j} \cdot \rho_{j,k}}{\|\rho_{i,j}\|\|\rho_{j,k}\|}\right]$$

and dihedral angles τ

$$\mathbf{v} = \rho_{i,j} \times \rho_{k,j}$$

$$\mathbf{w} = \rho_{k,j} \times \rho_{k,\ell}$$

$$\tau = \cos^{-1}\left[\frac{\mathbf{v} \cdot \mathbf{w}}{\|\mathbf{v}\|\|\mathbf{w}\|}\right]$$

we may compute the energies using these formulas:

2-Body Terms

$$\epsilon_{Bond} = C_B(\rho - \rho_o)^2 \quad \text{Bond}$$

$$\epsilon_{L-J} = \frac{A}{\rho^{12}} - \frac{B}{\rho^6} \quad \text{Lennard-Jones}$$

$$\epsilon_{COUL} = \frac{1}{4\pi\epsilon_o}\frac{1}{\rho} \quad \text{Coulomb}$$

3-Body Terms

$$\epsilon_{Angle} = C_A(\theta - \theta_o)^2 \quad \text{Valence angle}$$

4-Body Terms

```
[ atom ]   (coordinates of BI)
    # TYPE       X        Y        Z     Q
    1  CH2     0.153    0.000    0.100    0
    2  CH2     0.076    0.133    0.000    0
    3  CH2    -0.077    0.133    0.000    0
    4  CH2    -0.153    0.000    0.000    0
    5  CH2    -0.076   -0.133    0.000    0
    6  CH2     0.076   -0.133    0.100    0
[ bond  ]
    #  i   j     RHO        CB
    1  1   2  0.15300    334720.
    1  1   2  0.15300    334720.
    2  2   3  0.15300    334720.
    3  3   4  0.15300    334720.
    4  4   5  0.15300    334720.
    5  5   6  0.15300    334720.
    6  6   1  0.15300    334720
[ angle ]
    #  i   j   k    THETA      CT
    1  1   2   3    109.5    460.24
    1  2   3   4    109.5    460.24
    1  3   4   5    109.5    460.24
    1  4   5   6    109.5    460.24
    1  5   6   1    109.5    460.24
    1  6   1   2    109.5    460.24
[ dihedral ]
    #  i   j   k   l  DELTA   N    CD
    1  1   2   3   4      0   3  5.858
    2  2   3   4   5      0   3  5.858
    3  3   4   5   6      0   3  5.858
    4  4   5   6   1      0   3  5.858
    5  5   6   1   2      0   3  5.858
    6  6   1   2   3      0   3  5.858
[ non-bonded ]
    #  i   j          A             B
    1  1   4  0.472360E-02   0.711450e-05
    2  2   5  0.472360E-02   0.711450e-05
    3  3   6  0.472360E-02   0.711450e-05
```

TABLE A.1

Cyclohexane Topology, united atoms

$$\epsilon_{Dihedral} = C_D(1 + \cos[n\tau - \delta])$$

where δ is a phase angle, n is a small integer.

In the cyclohexane example there are no charges; thus the energy is

$$\epsilon = \epsilon_{Bond} + \epsilon_{Lennard-Jones} + \epsilon_{Angle} + \epsilon_{Dihedral}$$

A "molecular topology" is a data set which allows one to compute this value as a function of the positions of all atoms in the molecule. The cyclohexane topology is simple to construct, so we give it in the accompanying Table A.1, taking published interaction constants from the GROMOS forcefield [28]. As usual we treat this molecule as a set of six "united atom methylene groups," i.e. 6 points, each representing the carbon-hydrogen group $,CH_2$. The geometry of the molecule is initiated as a more-or-less regular hexagon, perturbed by raising two opposing vertices to simulate the "boat" structure covered in the text. The bonded interaction is thus composed of 6 bonds, 6 valence angles, and 6 dihedral angles. The are no Coulomb terms; the non-bonded interaction is composed of three "1-4" interactions. Energies and forces for the model calculations in the text were made using this forcefield.

REFERENCES

[1] Iupac-iub commission on biochemical nomenclature. *Biochemistry*, 9:3174–3179, 1970.

[2] F. Alluffi-Pentini, V. Parisi, and F. Zirilli. Global optimization and stochastic differential equations. *J. Optim. Theory Appl.*, 47:1–16, 1985.

[3] J. A. Barker and D. Henderson. Dynamics of proteins and nucleic acids. *Rev. Mod. Phys.*, 48:587–671, 1976.

[4] H. J. C. Berendsen, J. P. M. Postma, W. F. van Gunsteren, A. DiNola, and J. R. Haak. Molecular dynamics with coupling to an external bath. *J. Chem. Phys.*, 81:3684–3690, 1984.

[5] R. A. P Butler and E. E. Slaminka. An evaluation of the sniffer global optimization algorithm using standard test functions. *J. Comp. Phys.*, 99:28–32, 1992.

[6] R. Car and M. Parinello. Unified approach for molecular dynamics and density-functional theory. *Phys. Rev. Lett.*, 55:2471–2474, 1985.

[7] L. C. W. Dixon and G. P. Szego, editors. *Towards Global Optimization, Volume 2.* Elsevier, New York, 1978.

[8] R. A. Donnelly and J. W. Rogers, Jr. A discrete search technique for global optimization. *Intl. J. Quantum Chem.*, 22:507–513, 1988.

[9] A. O. Griewank. Generalized descent for global optimization. *J. Optim. Theory Appl.*, 34:11–39, 1981.

[10] T.S. Harvey and A.J. Winkinson andI.D. Campbell. The solution structure of human transforming growth factor α. *Eur. J. Biochem.*, 989:555–562, 1991.

[11] M.R. Hoare and J. McInnes. Morphology and statistical statics of simple microclusters. *Adv. Phys.*, 32:791, 1983.

[12] M.R. Hoare and P. Pal. *Adv. Phys.*, 20:161, 1971.

[13] M.R. Hoare and P. Pal. *Adv. Phys.*, 24:645, 1975.

[14] S. Kirkpatrick, Jr. G. D. Gelatt, and M. P. Vecchi. Optimization by simulated annealing. *Science*, 220:671–680, 1983.

[15] J. A. McCammon and S. C. Harvey. *Dynamics of Proteins and Nucleic Acids.* Cambridge University Press, New York, 1987.

[16] J.A. Northby. Structure and binding of lennard-jones clusters. *J. Chem. Phys.*, 87:6166, 1987.

[17] M.L. Papay. Glide program optimization results. Technical report, TRW Defense Systems Group, San Bernadino, CA, 1989. Unpublished.

[18] R. Stephen Berry Ralph E. Kunz. Statistical interpretation of topographies adn dynamics of multidimensional potentials. *J. Chem. Phys.*, 103:1904, 1995.

[19] J. W. Rogers, Jr and R. A. Donnelly. A search technique for global optimization in a chaotic environment. *J. Optim. Theory Appl.*, 61:111–121, 1989.

[20] J. W. Rogers, Jr and R. A. Donnelly. Potential transformation methods for large-scale global optimization. *SIAM J. Opt.*, 5:871, 1995.

[21] J.P. Ryckhaert, G. Ciccotti, and H. J. C. Berendsen. Time-dependent constraints. *J. Comput. Phys.*, 23:327, 1977.

[22] Panos M. Pardalos David Shalloway and Guoliang Xue. Optimization methods for computing global minima of nonconvex potential energy functions. *J. Global Opt.*, 4:117, 1994.

[23] Thomas Coleman David Shalloway and Zhijun Wu. Build-up algorithms for global energy minimization of molecular clusters using effective energy simulated annealing. *J. Global Opt.*, 4:171, 1994.

[24] E. E. Slaminka and K. D. Woerner. Central configurations and a theorem of palmore. *Celestial Mech. Dyn. Astron.*, 48:347–355, 1990.

[25] J. A. Synman and L. P. Fatti. A multi-start global minimization algorithm with dynamic search trajectories. *J. Optim. Theory Appl.*, 54:121–141, 1987.

[26] A. A. Törn. A search clustering approach to global optimization. In L. C. W. Dixon and G. P. Szego, editors, *Towards Global Optimization, Volume 2*, New York, 1978. Elsevier.

[27] C.J. Tsai and K.D. Jordan. Use of the eigenmode method to locate the stationary points of the potential energy surfaces of selected argon and water clusters. *J. Phys. CHem.*, 93:11227, 1993.

[28] David van der Spoel, Rudi van Drunen, and Herman J.C. Berendsen. Groningen machine for simulating chemistry, gromax user manual. Technical report, BIO-SON Research Institute, Nijenborgh 4, NL 9747 AG, Groningen, Netherlands, 1994.

[29] D. Vanderbuilt and S. G. Louie. Optimization by simulated annealing. *J. Comput. Physics*, 56:259, 1984.

[30] Guoliang Xue. Improvement of the northby algorithm for molecular conformation: Better solutions. *J. Global Opt.*, 4:425, 1994.

[31] Guoliang Xue. Molecular conformation on the cm-5 by parallel two-level simulated annealing. *J. Global Opt.*, 4:187, 1994.

MULTISPACE SEARCH FOR PROTEIN FOLDING

JUN GU[1], BIN DU[1], AND PANOS PARDALOS[2]

Abstract. Molecular energy minimization is one of the most challenging, unsolved problems in molecular biophysics. Due to numerous local minima in the search space, a traditional optimization method has a tendency to get stuck at some local minimum points. In this paper, for Lennard-Jones clusters, we give a multispace search algorithm for molecular energy minimization. Multispace search interplays structural operations in conjunction with the existing optimization methods. Structural operations dynamically construct a sequence of intermediate lattice structures by changing the original lattice structure. Each intermediate lattice structure is then optimized by the traditional optimization methods. Structural lattice operations disturb the environment of forming local minima, which makes multispace search a very natural approach to molecular energy minimization. We compare multispace approach with traditional optimization techniques for molecular energy minimization problems.

Key words. Molecular energy minimization, optimization algorithm, Lennard-Jones cluster, multispace search, protein folding, local search.

1. Introduction. Molecular energy minimization is one of the most challenging, unsolved problems in molecular biophysics [1,26]. In recent years, this problem has attracted much attention from mathematicians and computer scientists because the problem is itself a benchmark for the existing optimization methods. One of the fundamental problems in molecular energy minimization is to minimize the Lennard-Jones (LJ) potential function [3], which can be stated as follows:

Given a cluster of n atoms interacting with two-body central forces, our goal is to find their configuration possessing the globally minimum total potential energy.

The total Lennard-Jones potential energy function of n atoms (a_1, a_2, ..., a_n) can be defined as follows:

$$E_n = \sum_{1 \le i \le j \le n} V_{LJ}(\| a_i - a_j \|),$$

where $\| a_i - a_j \|$ is the Euclidean norm distance and $V_{LJ}(r)$ is the Lennard-Jones potential function of two atoms. It is defined as

---

[1] Department of Electrical and Computer Engineering, University of Calgary, Calgary, Alberta T2N 1N4, Canada, gu@enel.ucalgary.ca, and Department of Computer Science, The Hong Kong University of Science and Technology, Clear Water Bay, Kowloon, Hong Kong, gu@cs.ust.hk. This work was supported in part by NSERC Strategic Grant MEF0045793 and NSERC Research Grant OGP0046423 and is presently supported in part by NSERC Strategic Grant STR0167029 and the Federal Micronet Research Grant.

[2] Center for Applied Optimization and ISE Department, University of Florida, Gainesville, FL 32611, U.S.A. Research of the third author was partially supported by DIMACS (DIMACS is a National Science Foundation Science and Technology Center, funded under contract - STC - 91- 19999) and in part by NSF grant BIR-9505919.

$$V_{LJ}(r) = \frac{1}{r^{12}} - \frac{2}{r^6}.$$

The Lennard-Jones function is widely used in molecular energy minimization. It provides a good model for the molecular conformational energy that governs the behavior of simple physical systems — molecular clusters of chemically inert atoms (e.g., argon) [2]. The difficulty of the LJ cluster problem arises from the fact that it is a global optimization problem with an exponential number of local minima [15,16]. M.R. Hoare [14] claimed that the number of local minimizers of an n-atom cluster grows as fast as the function $O(e^{n^2})$. L.T. Wille and J. Vennik [30] showed that the complexity of determining the global minimum energy of a cluster of particles interacting via two-body forces belongs to the class of NP-hard problems.

Much research has been conducted to solve the LJ cluster energy minimization problem. In this research, for Lennard-Jones clusters, we give a multispace search algorithm for molecular energy minimization. The algorithm dynamically interplays structural operations in conjunction with traditional optimization algorithms for molecular energy minimization. Experimental results indicate that structural operations are effective to handle the pathological behavior of local minima.

The rest of the paper is organized as follows: In the next section, we will briefly overview some existing molecular energy minimization methods for Lennard-Jones clusters. Section 3 describes the basic ideas of multispace search for solving optimization problems. A multispace search algorithm for molecular energy minimization is introduced in Section 4. The experimental results of the multispace search algorithm for Lennard-Jones energy minimization are shown in Section 5. We will compare our algorithm with other algorithms for Lennard-Jones energy minimization problem. Section 6 concludes this paper.

2. Previous work. Many optimization methods have been developed for molecular energy minimization. These include lattice based search [22,32], simulated annealing and genetic algorithms [29], diffusion equation method [19], effective energy methods [27], global optimization [23], and packet annealing [28]. For a detailed survey of the optimization algorithms for molecular energy minimization, we refer readers to some recent survey papers [24,25].

Lattice based search has achieved great success in the minimization of Lennard-Jones energy function. J.A. Northby [22] developed an efficient lattice based search procedure. In his approach, he proposed and used IC and FC lattices [22] as the base lattice structures for constructing the initial Lennard-Jones clusters. The initial lattice configurations are allowed to relax freely on the Lennard-Jones pair potential to the adjacent energy minimum. Northby reported solutions for Lennard-Jones clusters of size

$13 \leq n \leq 150$. These were believed to be the most tightly bound configuration [22].

Later, Xue reduced the complexity of the Northby algorithm from $O(n^{\frac{5}{3}})$ to $O(n^{\frac{2}{3}})$ by introducing simple data structures [31]. He gave the detailed construction methods for the IC and FC lattice structures and improved the Northby algorithm by relaxing every lattice local minimizer. Although his algorithm requires more computing time, Xue found lower energy configurations for $n = 65, 66, 75, 76, 77$, and 134.

Simulated annealing is a general optimization approach proposed by Kirkpatrick et al. [18]. Wille used simulated annealing and found a new minimum solution for a 24-atom Lennard-Jones cluster [29]. Further, Xue proposed a two-level simulated annealing algorithm that worked well for large Lennard-Jones clusters [32]. Coleman, Shalloway, and Wu used simulated annealing and found a better structure for the 72-atom Lennard-Jones cluster [2].

Genetic algorithms are a class of search algorithms based on the mechanics of natural selection and natural genetics. Judson et al. implemented both simulated annealing and genetic algorithm for Lennard-Jones energy minimization [17]. Experimental results showed that simulated annealing and genetic algorithms have complementary strengths which implies that a hybrid GA-SA method would be more efficient than either one alone.

In addition to the above approaches, special techniques based on spatial averaging [19,27,28] and other optimization methods [21] have been proposed.

Recently a general optimization approach, *multispace search*, has been developed for optimization problem solving [4,9,10]. Multispace search can be applied to general protein folding problems. We have used multispace search to the Lennard-Jones energy minimization problem. The algorithm interplays structural operations in conjunction with the existing optimization algorithms. During the optimization process, structural operations dynamically construct a sequence of intermediate lattice structures by changing the original lattice structure. Each intermediate lattice structure is optimized, producing an improved intermediate solution. Eventually the intermediate lattice structure approaches the given problem structure. Structural operations change the intermediate lattice structures, disturbing the environment of forming permanent local minima (the local minima of an intermediate lattice structure will be changed when structured operations build another new intermediate lattice structure). This makes multispace search a very natural approach to handle the molecular energy minimization problem.

3. Optimization by multispace search. In this section, we briefly introduce the basic ideas of multispace search.

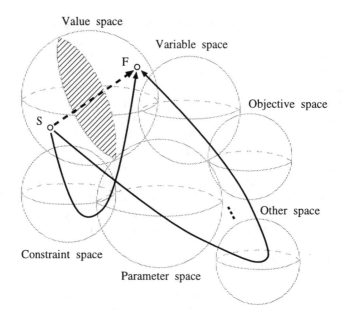

Fig. 3.1. *In the value space, a traditional search process (dashed line) cannot pass a "wall" of high cost search states (hatched region). It fails to reach the final solution state, F. A multispace search process (solid lines) scrambles across different search spaces. It could bypass this "wall" through the other search spaces.*

3.1. Multispace search. Many search and optimization methods have been developed in combinatorial optimization, operations research, artificial intelligence, neural networks, genetic algorithms, and evolution programming. An optimization algorithm seeks a value assignment to variables such that all the constraints are satisfied and the performance objective is optimized. The algorithm operates by changing values to the variables in the value space. Because value changing does not affect the problem structure and the search space, it is difficult for a value search algorithm to handle the pathological behavior of local minima.

Multispace search is a new optimization approach developed in recent years [4,9,10]. The idea of multispace search was derived from principles of nonequilibrium thermodynamic evolution that structural changes are more fundamental than quantitative changes and evolution depends on the growth of new structure in biological system rather than just information transmission. A search process resembles the evolution process and the structural operations are important to improve the performance of traditional value search methods [4,9,10].

In multispace search, any active component related to the given problem structure can be manipulated, and thus, be formulated as an independent search space. For a given optimization problem, for its variables, values, constraints, objective functions, and key parameters (that affect the problem structure),[1] we define the variable space, the value space (i.e., the traditional search space), the constraint space, the objective function space, the parameter space, and other search spaces, respectively. The totality of all the search spaces constitutes a *multispace*.

The basic idea of multispace search is simple. Instead of being restricted in the value space, the multispace is taken as the search space. In the multispace, components other than value can be manipulated and optimized as well. During the search, a multispace search algorithm not only alters values in the value space; as shown in Figure 3.1, it also walks across the variable space and other active spaces, dynamically changes the problem structure in terms of variables, parameters, and other components, and systematically *constructs* a sequence of structured, intermediate problem instances. Each intermediate problem instance is solved by an optimization algorithm and the solution found is used as the initial solution to the next intermediate problem instance. By interplaying value optimization with structured operations, multispace search incrementally constructs the final solution to the search problem through a sequence of structured intermediate problem instances. Only at the *last* moment of the search, the reconstructed problem structure approaches the original problem structure, and thus the final value assignment represents the solution to the given search problem.

Because structural operations (related to variable, constraint, objective function, and parameter changes) introduce new structural information, a multispace search method is more effective to handle difficult optimization problems [4,9,12,13].

Multispace search algorithm combines traditional optimization algorithms with structural multispace operations. A typical multispace search algorithm is shown in Figure 3.2 [4,9,10]. It consists of the following three stages:

Initialization

At the beginning of search, for the given problem instance, an initial problem instance — which is usually different from the given problem instance — is designed. This initial problem instance will approach the given problem instance through a number of the altered, intermediate problem instances in a finite number of search phases.

A *scrambling schedule*, or *scrambling scheme*, specifies a sequence of events as *when* to make *what* structural changes to the intermediate problem instances in *which* active search space. During each search phase, the intermediate problem instance will be solved by a value search algorithm and the solution obtained will be passed to the next search phase as determined by the scrambling schedule. The last event in the schedule returns

---

[1] See an example in [12].

procedure Multispace_Search (MS) ()
begin
    /* initialization */
    input the given problem instance;
    design initial problem instance and scrambling schedule;
    solve the initial problem instance;

    /* scrambling between the active search spaces */
    while (*schedule* has events) **do**
    begin
        enter an active search space according to the event;
        construct a new intermediate problem instance;
        /* traditional value search in the value space */
        begin
            enter the value search space;
            solve the intermediate problem instance with value search;
            transmit the intermediate solution to the next search phase;
        end;
        if the intermediate solution is not the *final* solution
            then optimize it in the next search phase;
        if the problem instance becomes the original problem instance
            then return the *final* solution;
    end;
end;

FIG. 3.2. **MS:** *A general multispace search algorithm.*

the intermediate problem instance to the original, given problem instance.

Search

This major search step involves two fundamental operations: a traditional value search and the structural reconfiguration of the intermediate problem instance during each individual search phase. According to the active event in the schedule, the search process enters a specified search space and performs structural operations to the intermediate problem structures, followed by a traditional value search that optimizes the constructed intermediate problem instance. The resulting intermediate solution is then used as the initial problem instance to the next phase of multispace search.

The major structural operations in multispace search include multispace scrambling [4,5,9], simulated evolution [4,5,11,12], extradimension transition (e.g., air bridge, real dimension, and extra dimension) [4,7,8,13], search space smoothing [4,5,12], preprocessors (e.g., compression, partitioning, reorganization, and semi-processing) [4,13], and perturbations (e.g., jumping, tunneling, climbing, annealing, and indexing) [4,7,8,18].

Final state

At the *last* moment of multispace search, a dynamically constructed, intermediate problem structure must be replaced by the original problem structure. This results in the *final* search state. At this state, any value assignment to the variables gives the *final solution* to the given search problem.

3.2. Initial state. An important initial state for multispace search is the *uniform* state [4,5,10].

¿From nonequilibrium thermodynamic evolution, it is evident that symmetrical and asymmetrical interactions are two antagonistic processes operating simultaneously over the course of succession. At the beginning, homogeneity among species dominates the evolution. This suggests that a natural search process should initially start with a *relatively homogeneous state with almost symmetrical configurations*.

A typical homogeneous state is a *uniform state* where the initial structure of the search problem is completely symmetrical. In a uniform search state, there are no local minima in the search space. Any solution point in the search space gives the same performance figure. For graph based optimization problems, two typical uniform initial states are a *null graph* and a *complete graph* [5,11]. For path related optimization problems, three typical uniform initial states include the *average distances*, the *shortest distances*, and the *longest distances* [5,12]. Multispace search starting from the uniform initial search state improves many traditional search methods.

In the next section, we will give a multispace search algorithm for the molecular energy minimization problem.

4. A multispace search algorithm for molecular energy minimization. In this section, we give a multispace search algorithm for an n-atom LJ cluster molecular energy minimization problem.

In the following discussion, let:

- e_i be the energy of the ith atom. e_i is the sum of LJ potential energy between the ith atom and the other atoms:

$$e_i = \sum_{1 \leq j \leq n, j \neq i} V_{LJ}(\| a_i - a_j \|).$$

- E be the total energy of the n-atom LJ cluster:

$$E = \frac{1}{2} \sum_{i=1}^{n} e_i.$$

- C be the *core* of the largest IC lattice which contains fewer than n discrete lattice points. An IC lattice core may contain several spherical layers. If no confusion arises, we will use C to denote the index set of the lattice points on the IC core.

- S be the outer lattice surface covering the IC core. If no confusion arises, we will use S to denote the index set of the lattice points on the surface.
- I be a joint set of C and S, i.e., $I = C \cup S$ and $|I| = |C| + |S|$. In this model, the number of the initial lattice points generated on the IC lattice is greater or equal to the number of atoms in the LJ cluster, i.e., $|I| \geq n$.
- S_f and S_v be two subsets of S, representing the *filled* lattice positions and *vacant* lattice positions, on the surface S, respectively. In this model, $|S_f| = n - |C|$ [32] and $|S_v| = |I| - n$.

4.1. The algorithm. Based on the reconstruction of molecular lattice structures, in Figure 4.1, we give a multispace search algorithm for molecular energy minimization (*MEM*). The *MEM* algorithm is divided into three major procedures, i.e., an initialization procedure, a local search procedure, and a scrambling search procedure.

> **procedure** MEM()
> **begin**
> /* initialization */
> *complete_lattice* := generate_a_complete_IC_lattice();
> *problem_lattice* := remove_extra_surface_atoms(*complete_lattice*);
> E := compute_LJ_energy(*problem_lattice*);
>
> /* local search */
> *lattice* = local_search(*problem_lattice*);
>
> /* scrambling search */
> *optimal_lattice* = scrambling_search(*lattice*);
>
> **return** *optimal_lattice*;
> **end;**

FIG. 4.1. *A multispace search algorithm for molecular energy minimization (MEM).*

Initialization

Initially, procedure *generate_a_complete_IC_lattice*() generates the smallest IC lattice, I, that contains at least n atoms ($|I| \geq n$). The surface layer of I is S. Procedure *remove_extra_surface_atoms*() removes $|I| - n$ extra atoms from S, producing a lattice structure according to the given problem specification. Every time when an atom with the maximum energy is removed from the surface, the size of set S_v is incremented by one and the size of set S_f is reduced by one. Procedure *compute_LJ_energy*()

calculates the energy of each atom and then evaluates the total amount of energy of all the atoms on the given LJ lattice structure.

Local Search

The local search procedure contains mainly two parts, i.e., a discrete local search procedure that swaps atoms in the occupied set (S_f) and in the unoccupied set (S_v), reducing the LJ cluster energy. Following discrete local search, a continuous relaxation is performed by a pattern search method which further minimizes the LJ cluster energy.

Scrambling Search

Scrambling search procedure interplays a two-stage scrambling optimization procedure that removes/adds/expands some atom(s) from/to/in the lattice structure, followed by a pattern search procedure that relaxes the lattice structure and produces the lower molecular energy.

In the following, we will discuss the major procedures in the multispace search algorithm in detail.

4.2. The local search procedure. The algorithmic description of the local search procedure is shown in Figure 4.2.

Discrete local search

The local search procedure is used to find a lattice structure with locally minimum LJ potential energy. For an atom in the ith lattice position, a_i, and an atom in the jth lattice position, a_j, if $e_i > e_j$, i.e., an atom in the ith lattice position possesses higher potential energy than in the jth lattice position, then the atom in the ith position will be moved to the jth position, followed by an update of memberships in sets S_f and S_v. This is carried out by procedure $perform\_swap(a_i, a_j)$. The local search process repeats to swap the atoms in two sets in a sequential order until no improvement occurs, i.e., a lattice structure with locally minimum energy is found.

¿From our experience, local search is very efficient in two aspects. First, at the beginning of the search, using a full assignment of lattice positions to all the atoms, it reduces down an exponential growth search space to a manageable one. Secondly, it searches for improvement within its local neighborhood as long as there is some improvement. A major weakness of local search is that it has a tendency of getting stuck at a locally optimum configuration (i.e., a local minimum point).

In the *MEM* algorithm, we have used structural operations to disturb the environment of forming local minima. So, in this *local_search()* procedure, we only use a simple local handler (Figure 4.3) to overcome the

procedure local_search(*problem_lattice*)
begin
    lattice := *problem_lattice*;
    E_{min} := compute_LJ_energy(*lattice*);
    k_{max} := 40;

    k := 0;
    while $k < k_{max}$ **do**
    begin
        /* discrete local search */
        for the ith atom a_i in S_f **do**
            for the jth atom a_j in S_v **do**
                if $e_i > e_j$ **then**
                    problem_lattice := perform_swap(a_i, a_j);

        /* continuous relaxation by pattern search method */
        problem_lattice := min_by_pattern_search(*problem_lattice*);
        E := compute_LJ_energy(*problem_lattice*);
        if $E < E_{min}$ **then**
            lattice := *problem_lattice*; E_{min} := E;
        else
            problem_lattice := *lattice*;
            local_handler(*problem_lattice*);
            k := $k + 1$;
    end;

    /* return the improved lattice structure */
    return *lattice*;
end;

FIG. 4.2. *The local search procedure.*

weakness of local search. Procedure *local_handler*() randomly swaps a pair of atoms in S_f and S_v. The major iteration loop in the local search repeats some 40 times before the *MEM* algorithm enters into the scrambling search procedure. That is, constant k_{max} is initialized to be 40. The locally minimum lattice obtained from local search procedure is used as the initial lattice structure for scrambling search.

Continuous relaxation by pattern search method

Local search is efficient for discrete optimization [6,7]. As the first processing step, local search is used to minimizing the molecular energy of the Lennard-Jones cluster. Local search produces a set of *discrete* lattice

```
procedure local_handler(problem_lattice)
begin
    for a random pair of atoms in S_f and S_v do
    begin
        problem_lattice := perform_swap(a_i, a_j);
        E := compute_LJ_energy(problem_lattice);
    end;
end;
```

FIG. 4.3. *A simple local handler.*

positions which are close to the lattice points generating the minimum LJ energy. Local search, however, will not be able to locate the exact position with the minimum LJ energy since the problem is itself a global optimization problem in the real space. After each discrete local search procedure, we use a continuous relaxation method, i.e., pattern search, to further search for an accurate *real* position that gives lower LJ potential energy.

Pattern search method, developed by Hooke and Jeeves in 1961, is similar to the coordinate ascent method. In pattern search, each coordinate is searched cyclically in order to identify the best local descent direction. Then the search process is accelerated along this favorable descent direction. Since pattern search is simple, the entire process can be executed fairly efficiently in linear time. Unlike optimization methods requiring partial derivatives, pattern search method does not require the calculation of the function derivatives. So it is simple to implement.

In our pattern search algorithm, we use the lattice structure from the discrete local search or the scrambling search as the initial lattice structure. For a chosen step d, for each variable x_i in the kth iteration, we cyclically test if $E(x_i^k \pm d)$ is less than $E(x_i^k)$. If it is, we have

$$\acute{x}_i^{k+1} := x_i^k \pm d,$$

otherwise,

$$\acute{x}_i^{k+1} := x_i^k.$$

This produces a vector $\acute{\mathbf{x}}^{k+1}$. ¿From $\acute{\mathbf{x}}^{k+1}$, we can obtain an accelerated vector, $\ddot{\mathbf{x}}^{k+1}$, along the $\mathbf{d}$ direction:

$$\ddot{\mathbf{x}}^{k+1} := 2\,\acute{\mathbf{x}}^{k+1} - \mathbf{x}^k.$$

The next new solution vector, i.e., $\mathbf{x}^{k+1}$, is:

$$\mathbf{x}^{k+1} = \begin{cases} \acute{\mathbf{x}}^{k+1} & \text{if } E(\acute{\mathbf{x}}^{k+1}) < E(\ddot{\mathbf{x}}^{k+1}); \\ \ddot{\mathbf{x}}^{k+1} & \text{otherwise.} \end{cases}$$

The step d will be further reduced if the pattern search process is retarded. The process is terminated if reducing d does not have any effect, or the precision limit prohibits d from being further reduced.

4.3. Scrambling search procedure. The scrambling search procedure, which is the third major part of *MEM*, is shown in Figure 4.4. It consists of a two-stage scrambling optimization procedure followed by a continuous relaxation using the pattern search method.

Scrambling search with atom movement

Local search may work better at earlier stage of optimization. When the molecular energy is reduced to a lower value, it is difficult for local search to proceed due to the pathological behavior of local minima. We use extra, structural operations to help the local search procedure to proceed.

During the initialization procedure (Figure 4.1), $|I|$ atoms on the initial IC lattice structure were generated. $|I| - n$ redundant atoms were removed from the initial lattice structure and their lattice positions were kept in the S_v set. In the first stage of scrambling optimization, some structural operations, i.e., atom movement, are performed. Each time an atom is added to a lattice position in S_v, followed by a pattern search that reduces the molecular energy of this $(n + 1)$-atom LJ cluster. The LJ system now has one extra atom. Then, an atom with the maximum energy is identified and is removed from the LJ cluster, followed by a pattern search that minimizes the molecular energy of this n-atom LJ cluster. ¿From the above operations, if the molecular energy of the LJ cluster is reduced, the new lattice structure is kept and the minimum energy updated; otherwise, the previously optimal lattice structure is recovered.

Scrambling search with surface expansion

In the second stage of scrambling search, we perform structural operations to the lattice structure by expanding the lattice surface. This can be done by multiplying the position of each atom at the surface to the center of mass of the lattice configuration by a constant factor *scale*. In procedure *expand_lattice_surface*() (Figure 4.4), for each atom at the lattice surface, its coordinates (x, y, z) are expanded by multiplying the *scale*,

$$x = (x - X0) * scale + X0$$
$$y = (y - Y0) * scale + Y0$$
$$z = (z - Z0) * scale + Z0$$

where $(X0, Y0, Z0)$ are the coordinates of the mass center of the lattice and, from the experimental observations, the *scale* is set to be 1.1 and 1.7 in two different iterations.

5. Experimental results. The *MEM* algorithm for the minimization of the molecular energy of the Lennard-Jones cluster was written in C. The experiments were performed on SUN SPARC 5 or SPARC 20 workstations.

For Lennard-Jones clusters with sizes varying from 51 to 300, Table 5.1 gives computational results of the multispace search algorithm for the minimization of the molecular energy of the LJ clusters. In the table, the first column indicates the number of atoms, n, in the LJ clusters. The second, third, and fourth columns, which are labeled with E_0, E_{local}, and $E_{scramble}$, denote the molecular energy of the LJ clusters after the initialization, local search, and scrambling search procedures, respectively. The last column gives the computing results of Xue's algorithm for the same size Lennard-Jones clusters [31]. Xue's algorithm was implemented on a 64-bit Cray-XMP computer at the Minnesota Supercomputer Center.

¿From this table we can see that the multispace search algorithm found the best known molecular energies for most LJ clusters [32,31,20]. For LJ clusters having 88, 98, 107, 113, and 115 atoms, the multispace search algorithm found lower molecular energy values than those reported earlier. The coordinates of the 115-atom LJ cluster are listed in Table 5.2.

The minimization of molecular energy for the LJ clusters is a difficult global optimization problem. We observed that the difficulty does not only depend on the size of the LJ cluster, n. It also depends on how close the cluster size n is to the icosahedral "magic number" [20]. A larger distance between n and the magic number makes this problem harder.

The experimental results for icosahedral structures $IC_2 - IC_8$ are shown in Table 5.3. In this case, the outer surface of IC lattice is fully occupied. So it is relatively easy to find an optimization solution. Our multispace search algorithm found all the best known molecular energy values for the LJ clusters in reasonable CPU time.

6. Conclusion. For most molecular energy minimization problems, the energy function may be reduced rapidly at the early stage of optimization, showing an *easy phase* of progression. The *difficult phase* of the molecular energy minimization appears at the later stage of finding a globally optimum solution. During this stage, due to the existence of many local minima, it is difficult for most optimization algorithms to reduce the molecular energy function by even a very small percentage.

To cope with the exponential number of local minima, for Lennard-Jones clusters, we have developed a multispace search algorithm for molecular energy minimization. The algorithm interplays structural operations in conjunction with traditional optimization techniques. Structural lattice operations disturb the environment of forming local minima, which makes multispace search a natural approach to molecular energy minimization. Experimental results indicate that, in addition to achieving the best known results in most cases, for 88-atom, 98-atom, 107-atom, 113-atom, and 115-atom Lennard-Jones clusters, our algorithm found the new lattice structures whose molecular energies are lower than those found in the previous studies.

TABLE 5.1

Experimental results of the multispace search algorithm for molecular energy minimization of the Lennard-Jones clusters. E_0, E_{local}, and $E_{scramble}$ denote the LJ molecular energy after the initialization, local search, and scrambling search procedures, respectively.

n	E_0	E_{local}	$E_{scramble}$	Xue [31]
51	-248.550	-251.254	-251.254	-251.254
52	-255.480	-258.230	-258.230	-258.230
53	-262.411	-265.203	-265.203	-265.203
54	-269.374	-272.209	-272.209	-272.209
55	-276.367	-279.248	-279.248	-279.248
56	-280.589	-283.643	-283.643	-283.643
57	-284.918	-288.140	-288.343	-288.343
58	-289.894	-293.225	-294.378	-294.378
59	-294.912	-299.302	-299.738	-299.738
60	-300.290	-304.491	-305.875	-305.876
61	-305.430	-310.044	-312.009	-312.009
62	-310.827	-316.235	-317.354	-317.354
63	-316.118	-321.550	-323.490	-323.490
64	-322.408	-327.850	-329.620	-329.620
65	-328.955	-334.447	-334.918	-334.972
66	-335.112	-340.047	-340.047	-341.111
67	-341.300	-346.351	-346.351	-347.252
68	-346.518	-352.684	-352.684	-353.395
69	-352.736	-359.018	-359.018	-359.726
70	-359.184	-365.351	-365.351	-366.892
71	-365.379	-371.714	-373.350	-373.350
72	-370.731	-377.294	-377.294	-378.524
73	-376.963	-383.639	-383.639	-384.789
74	-383.434	-390.009	-390.009	-390.908
75	-388.833	-395.734	-395.734	-396.037
76	-395.232	-402.083	-402.084	-402.385
77	-401.807	-408.463	-408.463	-408.518
78	-408.007	-414.680	-414.681	-414.681
79	-413.365	-421.482	-421.482	-421.811
80	-419.600	-427.829	-427.829	-428.084
81	-426.075	-434.206	-434.206	-434.344
82	-431.521	-440.550	-440.550	-440.550

n	E_0	E_{local}	$E_{scramble}$	Xue [31]
83	-437.928	-446.924	-446.924	-446.924
84	-444.521	-452.657	-452.657	-452.657
85	-450.848	-459.011	-459.012	-459.056
86	-458.174	-465.384	-465.384	-465.384
87	-464.412	-472.098	-472.098	-472.098
88	-469.774	-478.935	-479.033	-478.935
89	-476.015	-486.054	-486.054	-486.054
90	-482.494	-492.434	-492.434	-492.434
91	-487.943	-498.811	-498.811	-498.811
92	-494.352	-505.185	-505.185	-505.185
93	-500.949	-510.878	-510.878	-510.878
94	-507.314	-517.264	-517.264	-517.264
95	-514.646	-523.640	-523.640	-523.640
96	-520.022	-529.879	-529.879	-529.879
97	-526.398	-536.681	-536.681	-536.681
98	-532.902	-543.547	-543.643	-543.547
99	-538.354	-550.666	-550.667	-550.667
100	-544.768	-557.040	-557.040	-557.040
101	-551.368	-563.411	-563.411	-563.411
102	-557.734	-569.278	-569.278	-569.278
103	-565.068	-575.659	-575.659	-575.659
104	-570.448	-582.038	-582.038	-582.038
105	-576.826	-588.266	-588.266	-588.267
106	-583.335	-595.061	-595.061	-595.061
107	-588.873	-601.912	-602.007	-601.912
108	-595.299	-609.033	-609.033	-609.033
109	-601.966	-615.411	-615.411	-615.411
110	-609.435	-621.788	-621.788	-621.788
111	-614.895	-628.068	-628.068	-628.068
112	-621.356	-634.520	-634.875	-634.875
113	-627.977	-640.969	-641.795	-641.700
114	-634.351	-646.913	-646.913	-648.833
115	-641.692	-655.636	-655.756	-655.636
116	-647.238	-662.809	-662.809	-662.809
117	-653.709	-668.283	-668.283	-668.283

n	E_0	E_{local}	$E_{scramble}$	Xue [31]
118	-660.396	-674.770	-674.770	-674.770
119	-667.869	-679.559	-679.559	-681.419
120	-674.250	-686.679	-686.679	-687.022
121	-681.114	-691.114	-691.114	-693.820
122	-687.180	-698.301	-698.301	-700.939
123	-693.657	-705.490	-705.490	-707.802
124	-700.358	-712.678	-712.678	-714.921
125	-707.838	-717.497	-717.497	-721.303
126	-713.407	-724.682	-724.682	-727.350
127	-719.976	-731.869	-731.869	-734.480
128	-726.844	-739.059	-739.059	-741.332
129	-734.409	-744.197	-744.197	-748.461
130	-751.005	-751.753	-751.753	-755.271
131	-748.547	-758.934	-758.934	-762.442
132	-756.088	-765.979	-766.026	-768.042
133	-763.738	-773.678	-773.753	-775.023
134	-771.496	-781.476	-781.476	-782.206
135	-780.254	-790.278	-790.278	-790.278
136	-787.326	-797.453	-797.453	-797.453
137	-794.401	-804.631	-804.631	-804.631
138	-801.479	-811.813	-811.813	-811.813
139	-808.557	-818.994	-818.994	-818.994
140	-815.175	-826.175	-826.175	-826.175
141	-822.716	-833.358	-833.359	-833.359
142	-829.794	-840.539	-840.539	-840.539
143	-836.875	-847.722	-847.722	-847.722
144	-843.957	-854.904	-854.904	-854.904
145	-851.038	-862.087	-862.087	-862.087
146	-858.122	-869.272	-869.273	-869.273
147	-865.209	-876.461	-876.461	-876.461
148	-869.676	-881.073	-881.073	-881.073
149	-875.142	-886.693	-886.693	-886.693
150	-880.426	-893.310	-893.310	-893.310
200	-1209.830	-1229.185	-1229.185	-1229.185
300	-1914.944	-1942.106	-1942.107	-1942.107

TABLE 5.2

The coordinates of an optimal configuration for 115-atom Lennard-Jones cluster. The molecular energy of this configuration is $E = -655.756$.

i	x_i	y_i	z_i	i	x_i	y_i	z_i
1	0.032	-0.023	-0.028	2	0.014	-0.010	0.901
3	0.024	-0.017	-0.955	4	0.858	-0.016	0.395
5	0.699	0.474	-0.439	6	0.268	0.770	0.394
7	-0.239	0.770	-0.427	8	-0.649	0.472	0.393
9	-0.806	-0.010	-0.427	10	-0.649	-0.492	0.394
11	-0.234	-0.810	-0.439	12	0.281	-0.810	0.395
13	0.713	-0.518	-0.448	14	-0.003	0.002	1.853
15	0.002	-0.001	-1.915	16	1.706	-0.000	0.845
17	1.381	1.002	-0.861	18	0.512	1.584	0.825
19	-0.516	1.584	-0.839	20	-1.342	0.975	0.821
21	-1.665	0.001	-0.839	22	-1.348	-0.976	0.825
23	-0.525	-1.622	-0.861	24	0.528	-1.622	0.845
25	-0.015	0.012	2.838	26	0.848	0.002	1.357
27	0.259	0.800	1.352	28	-0.677	0.492	1.348
29	-0.680	-0.493	1.351	30	0.261	-0.807	1.356
31	0.702	0.509	-1.409	32	-0.260	0.799	-1.377
33	-0.840	0.001	-1.377	34	-0.266	-0.824	-1.409
35	0.706	-0.512	-1.419	36	1.567	0.509	-0.007
37	1.108	0.800	0.838	38	1.134	-0.823	0.859
39	1.578	-0.512	-0.007	40	0.946	1.299	-0.012
41	0.427	1.299	-0.851	42	1.412	0.001	-0.879
43	-0.000	1.598	-0.008	44	-0.417	1.290	0.830
45	-0.936	1.290	-0.010	46	-1.099	0.799	-0.848
47	-1.515	0.492	-0.010	48	-1.355	-0.002	0.829
49	-1.520	-0.493	-0.008	50	-1.102	-0.807	-0.852
51	-0.942	-1.300	-0.012	52	-0.417	-1.300	0.838
53	0.001	-1.647	-0.008	54	0.437	-1.342	-0.879
55	0.975	-1.342	-0.007	56	-2.246	0.006	0.424
57	-1.818	1.321	-0.431	58	-0.700	2.135	0.424
59	0.682	2.151	-0.429	60	-1.834	-1.313	-0.429
61	-0.725	-2.140	0.441	62	-1.660	-1.803	0.424
63	1.812	1.352	0.441	64	0.512	1.625	1.802
65	-0.798	2.424	-1.268	66	-1.158	-0.807	-1.829
67	0.250	0.821	2.325	68	0.840	0.015	2.335

i	x_i	y_i	z_i	i	x_i	y_i	z_i
69	-0.882	0.019	-2.345	70	1.695	0.020	1.828
71	0.769	2.425	1.270	72	-0.698	0.508	2.320
73	-1.378	1.001	1.794	74	-0.703	-0.491	2.325
75	-1.387	-0.989	1.802	76	0.246	-0.803	2.335
77	-0.290	0.834	-2.345	78	1.111	0.829	1.822
79	1.110	-0.806	1.844	80	-0.444	-1.312	1.821
81	-0.544	1.631	-1.808	82	-1.392	0.006	1.807
83	-1.718	0.014	-1.808	84	0.410	1.351	-1.829
85	0.505	-1.617	1.828	86	-1.138	0.827	-1.820
87	-0.436	1.322	1.808	88	-1.984	-0.804	-1.276
89	-1.251	-2.122	-0.424	90	1.960	0.835	1.299
91	1.366	1.632	1.284	92	-2.055	1.493	1.261
93	-2.067	-1.481	1.269	94	-1.414	-1.618	-1.281
95	-1.129	-1.803	1.284	96	1.631	1.847	-0.424
97	1.202	2.137	0.424	98	1.102	1.846	-1.280
99	0.152	2.136	-1.276	100	-2.551	0.010	-1.268
101	-0.187	-2.121	1.298	102	0.249	2.435	0.422
103	-0.274	2.435	-0.424	104	-0.174	2.121	1.271
105	-1.115	1.811	1.268	106	-1.222	2.120	-0.426
107	-1.641	1.810	0.417	108	-1.389	1.624	-1.276
109	-1.973	0.819	-1.276	110	-2.229	1.001	0.417
111	-2.394	0.507	-0.426	112	-2.066	0.501	1.268
113	-2.070	-0.490	1.271	114	-2.400	-0.492	-0.424
115	-2.238	-0.989	0.422				

TABLE 5.3

Experimental results of the multispace search algorithm for molecular energy minimization of the Lennard-Jones clusters for IC lattice structures with 2 (IC_2) to 8 (IC_8) spherical layers. E_0, E_{local}, and $E_{scramble}$ denote the LJ molecular energy after the initialization, local search, and scrambling search procedures, respectively.

IC_i	n	E_0	E_{local}	$E_{scramble}$	Maier et al. [20]
IC_2	13	-44.020	-44.327	-44.327	-44.327
IC_3	55	-276.367	-279.248	-279.248	-279.248
IC_4	147	-865.209	-876.461	-876.461	-876.461
IC_5	309	-1978.958	-2007.219	-2007.219	-2007.219
IC_6	561	-3785.915	-3842.393	-3842.393	-3842.394
IC_7	923	-6454.332	-6552.721	-6552.722	-6552.723
IC_8	1415	-10152.44	-10308.89	-10308.89	-10308.89

procedure scrambling_search(*lattice*)
begin
    E_{min} := compute_LJ_energy(*lattice*);
    optimal_lattice := *lattice*;

    /* scrambling on lattice structure */
    while there is improvement **do**
    begin
        /* Stage 1: scrambling with atom movement*/
        for each atom in S_v **do**
        begin
            lattice := *lattice* + *atom*;
            E := min_by_pattern_search(*lattice*);
            /* remove one atom with the maximum energy on the lattice */
            lattice := *lattice* − *atom*$_{(Max\ e_i)}$;
            E := min_by_pattern_search(*lattice*);
            if $E < E_{min}$ **then** *optimal_lattice* := *lattice*; E_{min} := E;
            else *lattice* := *optimal_lattice*;
        end;

        /* Stage 2: scrambling by expansion */
        k_{max} := 2;
        k := 0;
        while $k < k_{max}$ **do**
        begin
            scale := $1.1 + 0.6 \times k$;
            lattice := expand_lattice_surface(*lattice*, *scale*);
            E := min_by_pattern_search(*lattice*);
            if $E < E_{min}$ **then** *optimal_lattice* := *lattice*; E_{min} := E;
            else *lattice* := *optimal_lattice*;
            k := $k + 1$;
        end;
    end;
    return *optimal_lattice*;
end;

FIG. 4.4. *The scrambling search procedure.*

REFERENCES

[1] H.S. Chan and K.A. Dill. The protein folding problem. *Physics Today*, pages 24–32, Feb. 1993.

[2] T. Coleman, D. Shalloway, and Z. Wu. A parallel build-up algorithm for global energy minimizations of molecular clusters using effective energy simulated annealing. *Journal of Global Optimization*, 4(2):171–185, Mar. 1994.

[3] I.Z. Fisher. *Statistical Theory of Liquids*. University of Chicago Press, 1964.

[4] J. Gu. Optimization by multispace search. Technical Report UCECE-TR-90-001, Dept. of Electrical and Computer Engineering, Univ. of Calgary, Jan. 1990.

[5] J. Gu. Optimization by simulated evolution. Technical Report UCECE-TR-90-004, Dept. of Electrical and Computer Engineering, Univ. of Calgary, Mar. 1990.

[6] J. Gu. Efficient local search for very large-scale satisfiability problem. *SIGART Bulletin*, 3(1):8–12, Jan. 1992, ACM Press.

[7] J. Gu. Local search for satisfiability (SAT) problem. *IEEE Trans. on Systems, Man, and Cybernetics*, 23(4):1108–1129, Jul. 1993, and 24(4):709, Apr. 1994.

[8] J. Gu. Global optimization for satisfiability (SAT) problem. *IEEE Trans. on Knowledge and Data Engineering*, 6(3):361–381, Jun. 1994.

[9] J. Gu. *Constraint-Based Search*. Cambridge University Press, New York, 1995.

[10] J. Gu. *Optimization by Multispace Search*. Kluwer Academic Publishers, Massachusetts, 1996.

[11] J. Gu and B. Du. Graph partitioning by simulated evolution. Technical Report UCECE-TR-92-001, Dept. of Electrical and Computer Engineering, Univ. of Calgary, Apr. 1992.

[12] J. Gu and X. Huang. Efficient local search with search space smoothing. *IEEE Trans. on Systems, Man, and Cybernetics*, 24(5):728–735, May 1994.

[13] J. Gu and R. Puri. Asynchronous circuit synthesis by Boolean satisfiability. *IEEE Transactions on CAD*, 14(8):961–973, Aug. 1995.

[14] M.R. Hoare. Structure and dynamics of simple microclusters. *Advances in Chemical Physics*, 40:49–135, 1979.

[15] R. Horst and P.M. Pardalos (Editors). *Handbook of Global Optimization*. Kluwer Academic Publishers, 1995.

[16] R. Horst, P.M. Pardalos, and N.V. Thoai. *Introduction to Global Optimization*. Kluwer Academic Publishers, 1995.

[17] R.S. Judson, M.E. Colvin, J.C. Meza, A. Huffer, and D. Gutierrez. Do intelligent configuration search techniques outperform random search for large molecules? Technical Report SAND91-8740, Sandia National Laboratories, Center for Computational Engineering, Livermore, CA, Dec. 1991.

[18] S. Kirkpatrick, C.D. Gelat, and M.P. Vecchi. Optimization by simulated annealing. *Science*, 220:671–680, 1983.

[19] J. Kostrowicki, L. Piela, B.J.Cherayil, and H.A. Scheraga. Performance of the diffusion equation method in searches for optimum structures of clusters of Lennard-Jones atoms. *Journal of Physical Chemistry*, 95:4113–4119, 1991.

[20] R.S. Maier, J.B. Rosen, and G.L. Xue. A discrete-continuous algorithm for molecular energy minimization. In *Proceedings of IEEE/ACM Supercomputing'91*, pages 778–786. IEEE Computer Society Press, 1991.

[21] C.D. Maranas and C.A. Floudas. Global minimum potential energy conformations of small molecules. *Journal of Global Optimization*, 4(2):135–170, Mar. 1994.

[22] J.A. Northby. Structure and binding of Lennard-Jones clusters : $13 \leq n \leq 147$. *Journal of Chemical Physics*, 87(10):6166–6177, Nov. 1987.

[23] P.M. Pardalos. On the passage from local to global in optimization. In *Mathematical Programming: State of the Art 1994* (J.R. Birge & K.G. Murty, Editors), The University of Michigan, pp. 220-247.

[24] P.M. Pardalos, D. Shalloway and G.L. Xue. Optimization methods for computing global minima of nonconvex potential energy functions. *Journal of Global*

Optimization, 4(2):117–133, Mar. 1994.

[25] P.M. Pardalos, D. Shalloway and G. Xue (Editors). Global Minimization of Non-convex Energy Functions: Molecular Conformation and Protein Folding. *DI-MACS Series Vol. 23, American Mathematical Society*, (1996).

[26] F.M. Richards. The protein folding problem. *Scientific American*, pages 54–63, Jan. 1991.

[27] D. Shalloway. Application of the renormalization group to deterministic global minimization of molecular conformation energy functions. *Journal of Global Optimization*, 2:283–311, 1992.

[28] D. Shalloway. *Packet Annealing: A Deterministic Method for global Minimization. In* Recent Advances in Global Optimization, pages 433–477. Princeton University Press, 1992.

[29] L.T. Wille. Minimum-energy configurations of atomic clusters: New results obtained by simulated annealing. *Chemical Physics Letters*, 133:405–410, 1987.

[30] L.T. Wille and J. Vennik. Computational complexity of the ground-state determination of atomic clusters. *Journal of Physics A: Mathematic and General*, 18(8):L419–L422, June 1985.

[31] G.L. Xue. Improvement on the Northby algorithm for molecular conformation: Better solutions. *Journal of Global Optimization*, 4(4):425–440, 1994.

[32] G.L. Xue. Molecular conformation on the CM-5 by parallel two-level simulated annealing. *Journal of Global Optimization*, 4(2):187–208, 1994.

MODELING THE STRUCTURE OF ICE AS A PROBLEM IN GLOBAL MINIMIZATION

JAN HERMANS*

Abstract. The structure and properties of liquid water can be modeled with rigid H_2O molecules that interact via a simple potential function containing $1/r^6$ attractive, $1/r^{12}$ repulsive terms and $1/r$ (electrostatic) attractive and repulsive terms. In ice I, each water molecule is tetrahedrally surrounded by 4 other water molecules; one OH bond points along each O...O vector. Minimization problem 1: given the molecular packing of ice I, find the low-energy arrangement(s) of OH bonds. Minimization problem 2: find low-energy crystal structures of water, including the one of lowest energy.

1. Introduction. Important questions in structural biology await adequate minimization methods with large convergence range. Participants in the 1995 IMA workshop asked for well-defined problems that would have (some of) the mathematical complexity of problems in structural biology and could serve as commonly agreed upon benchmarks for testing progress in developing new minimization methods. Problems, such as the protein folding problem, are typically formulated in terms of minimization of an energy function; to be able to perform the repeated function evaluations needed in minimization procedures, simplified energy functions are used. Unfortunately, biophysicists disagree about the form these functions should have and about the parametrization of any particular form; indeed, it may be necessary to first solve the minimization problem before one can identify those functions which give good agreement with experimental observations!

Simplicity of form limits one to sums of pair potentials. An optimal representation uses interatomic pair potentials, and an all-atom representation; while more than one such force field exists (i.e., charmm, gromos, amber and others), these agree in form and differ only in detail. Even then, it is difficult to represent the solvent (water) and its effect on interatomic forces. To represent the solvent with explicit molecules introduces the question of how to represent the solvent's intrinsic disorder. Representation of the solvent as a dielectric continuum [3] promises good accuracy, but at a significant computational cost. When all atoms are represented, the number of terms in the energy function and the number of minima is large. New methods of summation of $1/r$ and $1/r^6$ energies can reduce the many-body problem to order $N \log N$ or order N complexity [2], [1]. Simplification to interactions between groups of atoms, rather than the atoms themselves, is a common way of dealing with the complexity. However, this introduces new arbitrary choices and difficult parametrization problems.

2. The problem. I suggest that a study of the structure of crystal forms of water, and in particular the ice I crystal form, can be a useful

* Department of Biochemistry and Biophysics, University of North Carolina, Chapel Hill, NC 27599-7260.

intermediate at current levels of achievement. A simple potential for inter-action between water molecules is available, with well-defined parameters. While this model has been developed for liquid water, we have recently found it to quite well represent some of the properties of ice [7].

The following two problems would seem to be of some interest. The most stable arrangement of water molecules at temperatures just below 273 K and at ordinary low pressure, is the ice I crystal structure. In this structure each water molecule is tetrahedrally surrounded by four other water molecules; one OH bond points along each O...O vector.

(1) The first problem is to assume the placement of the oxygen atoms on the crystal lattice of ice I, and to find the arrangement or arrangements of the hydrogen atoms of lowest energy.

(2) The second problem is to find low-energy crystal packings of water, including the one of lowest energy (which presumably corresponds to ice I), without prior knowledge of the arrangement.

3. The interatomic force field for water. Three simple models of the water molecule are available, which differ slightly in the values of the parameters [4] [6] [5]. The models have just three centers of interatomic interaction, which coincide with the positions of the three atomic point masses. The models are rigid, i.e., the OH bond lengths and the HH distance are fixed. Each atom has a partial charge, ϵ ($\epsilon_O = -2\epsilon_H$) from which the electrostatic part of the potential energy is computed; in addition, the oxygen atoms of two molecules interact via a Lennard-Jones potential. The potential energy, U is found as a sum over all pairs of molecules

$$
\begin{aligned}
U = \sum_{i=1}^{N} \sum_{j>i}^{N} \Bigg[&- A/r_{OiOj}^6 + B/r_{OiOj}^{12} \\
&+ (1/D)(\epsilon^2/r_{Hi1Hj1} \ldots - 2\epsilon^2/r_{OiHj1} \ldots + 4\epsilon^2/r_{OiOj}) \Bigg]
\end{aligned}
$$

(3.1)

where r represents interatomic distance.

The parameters are summarized in Table 1. (The unit of energy is kcal/mol, the unit of distance is Å and the unit of charge is the charge of the proton. With these units, the appropriate value of the factor $1/D$ is 332.0636.)

4. The structure of ice I. We have found it convenient to study the crystal structure of ice using a rectangular coordinate system in which a unit cell contains 8 water molecules. The distance between neighboring water molecules is $d = 2.766$ Å , and the dimensions of the cell are $a = 3d \sin \tau$, $b = 2d \sin \tau \sin 60°$ and $c = 8d/3$, with τ the tetrahedral angle. The first 4 oxygen atoms are placed at $(0,0,0)$, $(0,0,d)$, $(a/3,0,4d/3)$, and $(a/3, 0, 7d/3)$, and the positions of the other 4 are obtained by translating

Parameters for SPC, SPC/E and TIP3P water models.

parameter	SPC	SPCE	TIP3P
A	625.8	625.8	595.0
B	629,660	629,660	582,000
ϵ	0.410	0.4238	0.417
d_{OH}	1.00	1.00	0.9572 Å
θ_{HOH}	109.47	109.47	104.5°

these by $(a/2, b/2, 0)$. This volume should be replicated in order to obtain a system with many more independent molecules, for use in simulations with periodic boundary conditions. For example, a system containing 360 water molecules is obtained by using a box with volume $5a \times 3b \times 3c$.

REFERENCES

[1] J.A. BOARD, Z.S. HAKURA, W.D. ELLIOTT, W.J. BLANKE, D.C. GRAY, AND J.F. LEATHRUM, Scalable implementations of multipole-accelerated algorithms for molecular dynamics. In: Scalable high performance computing conference (SH-PCC '94). IEEE Computer Society, IEEE Computer Society Press, 1994, pp. 87–94.

[2] T. DARDEN, D. YORK, AND L. PEDERSEN, Particle mesh Ewald: An $N \log N$ method for Ewald sums in large systems. J. Chem. Phys. **98**, 10089–10092, 1993.

[3] K.S. SHARP AND B. HONIG, Electrostatic interactions in macromolecules: Theory and applications. Rev. Biophys. Biophys. Chem. **19**, 301–332, 1990.

[4] H.J.C. BERENDSEN, J.P.M. POSTMA, W.F. VAN GUNSTEREN, AND J. HERMANS, Interaction Models for Water in Relation to Protein Hydration. Jerusalem Symposia on Quantum Chemistry and Biochemistry, 1981, Reidel, Dordrecht, Holland, 331–342.

[5] H.J.C. BERENDSEN, J.R. GRIGERA, AND T.P. STRAATSMA. J. Phys. Chem. **91**, 6269–6271, 1987.

[6] W.L. JORGENSEN, J. CHANDRASEKHAR, J.D. MADURA, R.W. IMPEY, AND M.L. KLEIN. J. Chem. Phys. **79**, 926–935, 1983.

[7] S. KALAT AND J. HERMANS, Melting of ice in computer models. In progress, 1995.

NOVEL APPLICATIONS OF OPTIMIZATION TO MOLECULE DESIGN

J.C. MEZA*, T.D. PLANTENGA*, AND R.S. JUDSON*

Abstract. We present results from the application of two conformational search methods: genetic algorithms (GA) and parallel direct search methods for finding all of the low energy conformations of a molecule that are within a certain energy of the global minimum. Genetic algorithms are in a class of biologically motivated optimization methods that evolve a population of individuals where individuals who are more "fit" have a higher probability of surviving into subsequent generations. The parallel direct search method (PDS) is a type of pattern search method that uses an adaptive grid to search for minima. In addition, we present a technique for performing energy minimization based on using a constrained optimization method.

Key words. global optimization, constrained optimization, nonlinear programming, molecular conformation.

1. Introduction. An important goal of computational chemistry research is the design of molecules for specific applications. Examples of these types of applications occur in the development of enzymes for the removal of toxic wastes, the development of new catalysts for material processing, biosensor design and the design of new anti-cancer agents. Factors that must be taken into account include shape, size, electronic properties, and reactivity. For many physical and biological properties, the molecular conformation largely determines the final function, and this is the rationale for the development of a large number of conformation search methods.

The general approach is to search the conformation space of a molecule in order to find all energy minima within a prescribed energy range. The problem can be broken into two major parts: defining the energy function and finding efficient methods for performing the conformational search. In general, one can decompose the search into two phases. In the first phase, we are interested in performing a coarse but broad search. This stage generates a number of interesting conformations that can be used as starting guesses for the second phase, which is local energy minimization. The global search phase is conceptually the harder of the two because the size of the parameter space is so large. Additionally, local information about the surface rarely provides definitive clues regarding the location of the global minimum. Because it is difficult to exhaustively search the conformation space of any but the smallest molecules, a number of statistical heuristic methods have been developed [21,43]. These include pure random search, simulated annealing [45], Cartesian coordinate directed tweak [42], taboo search [11], parallel stochastic methods as in [7,8], genetic algo-

---

* Scientific Computing Department, MS 9214, Sandia National Laboratories, Livermore, CA 94551-0969, supported in part by the Department of Energy under contract DE-ACO4-94AL85000.

rithms [23,24,27] and direct search methods [29,32,33,46]. Non-stochastic methods have been developed, including Scheraga's diffusion method [25] and a class of branch-and-bound methods due to Floudas [30]. A discussion of many of the methods used for molecular conformation and protein folding can also be found in [37,38]. All of these methods can be made to work well on a selected set of molecules, but it is important to perform head-to-head tests between different methods to assess their relative strengths.

We are primarily concerned here with the efficient broad search of conformation space to generate all of the low energy conformations within a prescribed energy of a global minimum. In this sense, this work is similar in spirit to that of Saunders et al. [43]. In particular, we present a comparison of two search methods, the GA and PDS methods, for this problem. Genetic algorithms draw on a set of evolutionary metaphors including selection of fit individuals, mutation, and genetic crossover. PDS methods belong to a class of optimization algorithms developed by Dennis and Torczon [12] that can be viewed as multidirectional line search methods. These methods are robust, simple to implement, and easily parallelized. We note that neither of these methods has as its main goal that of finding all minima within a certain distance of the global minimum, but earlier work has indicated that both of these methods might be applicable to this problem.

In addition, we present a technique for computing local energy minimization based on a constrained optimization method. This method (originally described in [40]) is based on transforming an unconstrained optimization problem in torsion space into a constrained optimization problem using distance constraints that makes the energy minimization more tractable.

The paper is organized as follows. §2 describes the energy functional that we seek to minimize. §3 gives an outline of two conformational search methods used in this paper. §4 gives a description of the new constrained optimization method for performing energy minimization. In §5 we describe the test problems and give numerical results. §6 follows with an analysis of the results.

2. Potential energy equations. The conformational search problem can be stated as a problem of finding the molecular conformation that yields the lowest energy for a particular N-atom molecule, that is,

$$(2.1) \qquad\qquad \min \quad E(x, y, z),$$

where E is the energy of the molecule given its coordinates $x, y, z \in \mathbf{R}^N$. Using this formulation, the conformational search problem can be viewed as a global optimization problem. Unfortunately, because the total energy of a molecule depends on all atom-atom interactions, the number of possible low-energy configurations can grow exponentially with the number of atoms and has been estimated by Hoare to be on the order of $O(e^{N^2})$ for an N-atom molecule [19].

The energy that we wish to minimize can take many forms, but it is usually computed as a sum of terms that are functions of bond distances between two atoms, bond angles between three atoms, dihedral or torsion angles between four atoms, improper torsion angles, various non-bonded terms (Coulombic potentials, van der Waals potentials) and perhaps including solvent effects.

In this paper we will use the potential energy equations used in CCEMD (Center for Computational Engineering Molecular Dynamics) [22]. These equations correspond to the force field used in QUANTA/CHARMM19 [5,41] with modifications to comply with the force field in CHARMM22. Here we only present a brief statement of the major terms. The total energy is given by

$$(2.2) \qquad E = E_b + E_\theta + E_\phi + E_\omega + E_{LJ} + E_{el}.$$

The first four terms correspond to covalent bond interaction terms between atoms, while the last two terms correspond to nonbonded forces between any two atoms. The bonded energy term is defined by

$$(2.3) \qquad E_b = \sum_{i=1}^{N_{bond}} k_b^i (r^i - r_0^i)^2,$$

where k_b^i is the force constant for bond i, r^i is the bond distance, and r_0^i is the equilibrium bond distance. The bond angle energy term is given by

$$(2.4) \qquad E_\theta = \sum_{i=1}^{N_{angle}} k_\theta^i (\theta^i - \theta_0^i)^2,$$

where k_θ^i is the force constant for bond angle i, θ^i is the bond angle, and θ_0^i is the equilibrium bond angle. The dihedral angle energy term is

$$(2.5) \qquad E_\phi = \sum_{i=1}^{N_{dihedral}} \sum_j |k_i^j| - k_i^j \cos(n_i^j \phi^i),$$

where k_i^j is a force constant, n_i^j is an integer that can take on the values $1, 2, 3, 4, 6$ and ϕ^i is the dihedral angle. The improper torsion angle is defined as

$$(2.6) \qquad E_\omega = \sum_{i=1}^{N_{improper}} k_\omega^i (\omega^i - \omega_0^i)^2,$$

where k_ω^i is the force constant for improper torsion, ω^i is the torsion, and ω_0^i is the equilibrium torsion angle.

In addition to the bonded interactions, there are forces due to non-bonded interactions. The van der Waals term is usually taken to be of the form of a Lennard-Jones potential,

$$(2.7) \qquad E_{LJ} = \sum_{i \neq j} \left(\frac{A_{ij}}{r_{ij}^{12}} - \frac{B_{ij}}{r_{ij}^6} \right) \quad \text{sw}(r_{ij}),$$

where $A_{ij} = \varepsilon_{ij}\sigma_{ij}^{12}$, $B_{ij} = 2\varepsilon_{ij}\sigma_{ij}^6$, and $\varepsilon_{ij}, \sigma_{ij}$ depend on the atoms i and j. The term $\text{sw}(r_{ij})$ is a switching function that is used to cutoff the potential at long distances. A variety of switching functions may be used depending on the application (for details see [22]).

The final term due to the electrostatic potential is given by

$$(2.8) \qquad E_{el} = \sum \frac{q_i q_j}{4\pi\varepsilon_0 r_{ij}} \quad \text{sw}(r_{ij}),$$

where q_i and q_j are the charges on atoms i and j respectively and ε_0 is a dielectric constant.

3. Search methods. In this paper, we make the distinction between a *search* method and a *minimization* method for the conformation problem. By a search method we will mean any algorithm that is used to perform a coarse search of the parameter space to look for starting guesses for a gradient-based local minimization algorithm. In this sense, both genetic algorithms (GA) and parallel direct search (PDS) methods are good candidates for search methods since they can be used to quickly sample a large region of the parameter space. In addition, both of these methods are easily parallelized. This section describes the major features of the GA and PDS methods and how they are applied to the conformation problem.

3.1. Genetic algorithms. We present here only a brief introduction to our variant of the standard GA method [17]. The most important idea is that we work with a population of *individuals* that will interact through genetic operators to carry out an optimization process. An individual is specified by a *chromosome* that is a bit string of length N_c that can be decoded to give a set of physical parameters. In what follows, chromosome and bit string are synonymous. The function to be optimized, also called the fitness function, is used to rank the individual chromosomes. Optimization proceeds by generating populations whose individuals have increasingly higher fitness. An initial population of N_{pop} individuals is formed by choosing N_{pop} bit strings at random and evaluating each individual's fitness.

Conformations are represented as bit strings that code for the free torsion angles in the molecule. All bond distances and angles are held fixed. Each torsion is represented by n bits giving a resolution of $360/(2^n - 1)$

degrees. Typical values of n range from 6 to 12. If b is the Gray coded binary value of the angle, the value in degrees is $360b/(2^n - 1)$. The chromosome for the individual is constructed by concatenating the bit strings for the individual torsions. The three principal operators are selection of parents, crossover, and mutation.

In our selection operator, every individual in the top ranked 40% of the population has an equal chance of being selected for mating. All individuals in the bottom ranked 60% of the population are discarded. The fitness is the negative of the potential energy. The crossover operator takes a pair of parents and chooses a random cut point along the bit string. The chromosome of the first child is filled in with the bits to the left of the cut point from the first parent and the bits to the right of the cut point from the second parent. The second child gets the complementary bits from the two parents. Note that the crossover point is not restricted to lie on the boundaries separating dihedrals. Doing so restricts the search too much and leads to poorer performance by the GA. Notice however, that this introduces a subtle type of mutation because the dihedral that is disrupted by the cut point does not assume the angular value of either of the parents. By not restricting the cut point positions, we find that premature convergence occurs less often and that lower energies are found.

Finally the mutation operator acts by flipping bits in the binary chromosome. Each bit has a probability equal to R_m of being flipped from 1 to 0 or vice versa. Mutation rates are typically quite low, on the order of 0.04. An important detail is that the entire population is not regenerated at each generation. The top 10% of the old population is moved into the new population and all but the single best are subjected to the mutation operator, meaning that they are mutated with the same low probability as the rest of the members of the new population. We always use the *elitist* strategy in which the most fit individual in each generation is passed directly to the next without crossover or mutation. This ensures that the best individual is never lost, but continues to be available for mating. Note that this individual is transferred directly from generation i to $i + 1$ but also produces offspring that make up part of generation $i + 1$.

Additionally, during replication there is a small probability of a bit flip or mutation in a chromosome. This serves primarily to maintain diversity and prevent premature convergence that occurs when a single very fit individual takes over the entire population early in the evolutionary process. To bound the magnitude of the effect of mutations, the binary chromosomes are Gray coded [17]. An integer that is represented as a Gray coded binary number has the property that most single bit flips change the value of the integer by ± 1.

We have the ability to run multiple sub-populations simultaneously. At periodic intervals, these populations can communicate by passing the best individual from each population to each of the others.

During the crossover operations, a niching operation is used. As prospective new members of the population are created, they are compared to those already accepted, by measuring the Hamming distance. The Hamming distance is the fraction of bit positions that have different values in the two chromosomes. The prospective new member is rejected if it is too similar to ones already present. Initially, an individual must differ by 40% from every other individual (that is, no more than 60% of the bits in the two can by set the same.) As the population fills up, this criteria becomes too restrictive and it is slowly relaxed until the population is filled.

3.2. Direct search methods. Direct search methods belong to a class of optimization methods that do not compute derivatives. Examples of direct search methods are the Nelder-Mead Simplex method [35], Hooke and Jeeves' pattern search [20], the box method [4], and Dennis and Torczon's parallel direct search algorithm (PDS) [12].

The PDS algorithm can be described as follows. Starting from an initial simplex S_o, the function value at each of the vertices in S_o is computed and the vertex corresponding to the lowest function value, v_o, is determined. Using the underlying grid structure, the simplex S_o is rotated 180° about v_o and the function values at the vertices of this rotation simplex, S_r, are compared against v_o. If one of the vertices in the simplex S_r has a function value less than the function value corresponding to v_o, then an expansion step to form a new simplex, S_e, is attempted in which the size of S_r is expanded by some multiple, usually 2. The function values at the vertices of S_e are compared against the lowest function value found in S_r. If a lower function value is encountered, then S_e is accepted as the starting simplex for the next iteration; otherwise S_r is accepted for the next iteration. If no function value lower than the one corresponding to v_o is found in S_r, then a contraction simplex is created by reducing the size of S_o by some multiple, usually $1/2$ and is accepted for the next iteration.

Because PDS only uses function comparisons, it is easy to implement and use. Since the rotation, expansion, and contraction steps are all well-determined it is possible to determine ahead of time a set of grid points corresponding to the vertices of the simplices constructed from various combinations of rotations, expansions, and contractions. Given this set of grid points, called a search scheme, the PDS algorithm can compute the function values at all of these vertices in parallel and take the vertex corresponding to the lowest function value. An interesting consequence of this approach is that the PDS algorithm can jump out of local wells by using a large enough search scheme size. By varying the size of the search scheme one can therefore use the PDS algorithm as a means of efficiently generating conformations in a manner similar to GA and simulated annealing.

It is also worthwhile to contrast PDS with grid search methods. In a grid search method the grids are generated by starting with a fixed molecule and systematically varying one of the internal variables. This method works well for small molecules but becomes computationally prohibitive for larger molecules. The grid in PDS however is adaptive and will automatically change in response to the contours of the energy surface.

4. Energy minimization using constraints. The search method chooses candidate conformations for which the fitness must be evaluated. Fitness is usually some measure of the smallest potential energy that can be achieved from the candidate conformation. A simple measure is to compute the energy E for the conformation using equations (2.1)–(2.8). Much better results can be obtained by performing a local gradient minimization of the energy starting from the conformation chosen by the search method. However, gradient minimizations can be computationally expensive, especially for large molecules. We have chosen an alternative that makes a physically intuitive compromise. In this approach, the covalent structure of the molecule is fixed and local minimization is performed with only the dihedral angles as variables. This is also known as minimizing in torsion space [34]. The number of dihedral variables is much smaller than the number of atoms, so the minimization problem is easier to solve. In addition, fixing bond distances and bond angles eliminates many local energy minima that are similar in depth; hence, torsion space minima produce a simpler but still physically meaningful picture of the accessible conformations of a molecule.

It is possible to perform energy minimization in torsion space by expressing the potential energy and its analytic derivatives in terms of the dihedral variables. This approach was pioneered by Scheraga and coworkers in their ECEPP program [34] and has also been pursued by Gō et al. [2,44] and Abagyan et al. [31,1]. One of the difficulties with performing calculations in torsion space is that a complicated transformation of variables is required to go from Cartesian coordinates to a set of dihedral variables. The CHARMm potential energy model is most naturally expressed as a function of the Cartesian coordinates of each atom. Transforming this model to use dihedral variables and then computing analytic derivatives is quite complicated, usually requiring topological analysis of the molecule, definition of multiple local coordinate frames, and the use of matrix operators to link the local coordinate systems (see [16,2,31] for examples). In addition, the number of operations necessary to make a transformation is proportional to the square of the number of dihedral variables; thus, computational costs increase rapidly as larger molecules are examined.

Instead of transforming the potential energy to torsion space, we find a set of distance constraints between pairs of atoms that serves to restrict all molecular motion except rotation of specified dihedral angles. The distance constraints are simple quadratic functions in Cartesian coordinates. Thus, we minimize the usual potential energy function subject to a set of quadratic equality constraints, all in Cartesian coordinates. Our constrained problem is equivalent to minimizing in torsion space in the sense that we find the same set of local minima. However, the intermediate molecular conformations generated during our minimization are different. Our approach avoids the mathematical complexities associated with transforming to dihedral variables. We have a simple method of finding an appropriate set of distance constraints that is easily automated. Also, we are able to maintain variable sparsity in our constrained formulation, keeping the linear algebra costs manageable even for large molecules.

The chief advantage of our approach is that it makes numerical energy minimization in torsion space much simpler. Our method poses a minimization problem with nonlinear equality constraints in Cartesian coordinates instead of an unconstrained problem in internal coordinates. We have employed a code [26] based on sequential quadratic programming (SQP) methods that seems well suited to the constrained molecular mechanics problem. An SQP method treats the constraints explicitly and does not require the elimination of the dependent variables from the nonlinear constraints. The algorithm utilizes sparse linear algebra techniques to solve all subproblems and operates as a quasi-Newton or truncated Newton method with only first derivative information.

4.1. Dihedral variables and constraints. We follow Scheraga [34] and fix the covalent structure of the molecule except for the rotation of dihedral angles. This simplified model determines an approximate molecular conformation by changing dihedrals in response to non-bonded and dihedral angle forces. Dihedral angles are now the primary variables of the problem – atom positions are computed from knowledge of the dihedrals and the fixed bond distances and angles. An unconstrained minimization can be carried out in torsion space if we can calculate the gradient of the potential energy with respect to dihedral variables; that is, if we can transform the Cartesian force vector into dihedral coordinates. The transformation can be done analytically as in [2,31], but the equations are extremely complicated to derive and to program. We wish to show how the transformation of coordinates can be avoided using distance constraints between pairs of atoms.

Let us consider the simple example of ethane in Figure 1. It has a single dihedral angle, whose rotation is illustrated by the arrow. Usually, two planes are specified, such as H1-C1-C2 and C1-C2-H2, and the dihedral is defined as the angle between the two planes about the axis C1-C2. Now if all bond distances and angles in the ethane molecule are fixed, then the methane group on the left containing C1 and H1 is a rigid body that rotates about the C1-C2 axis. The methane group on the right is a similar rigid body. Thus, rotation about a dihedral can be characterized as the relative rotation of two rigid bodies about a common axis.

Figure 2 shows the ethane molecule with distance constraints (drawn as dashed and dotted lines) for the two methane groups. The distance constraints can be thought of as rigid bars or virtual bonds between atoms that restrict motion. The dashed constraints define a rigid five-point poly-

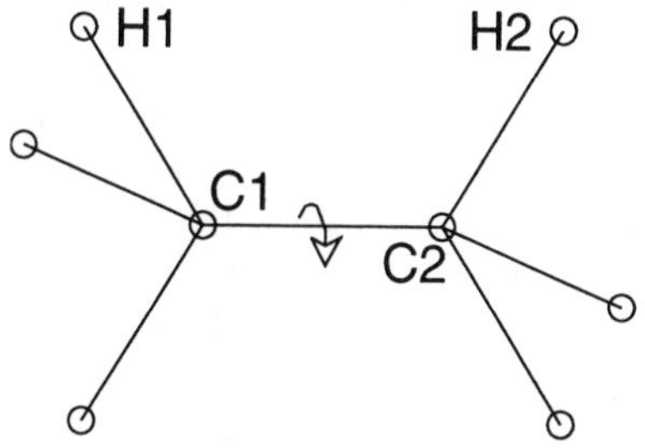

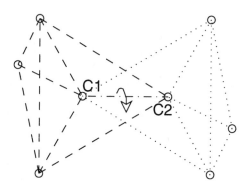

FIG. 1. *Dihedral rotational in ethane*

hedron on the left and the dotted constraints do the same on the right. The figure shows that these two rigid pieces are connected together in such a way that they can spin about the C1-C2 axis, but cannot otherwise move in relation to each other. Note in particular that the distance constraints for each rigid piece include the dihedral axis and reach across it.

FIG. 2. *Distance constraints allowing only dihedral rotation in ethane. There are 17 constraints restricting molecular motion.*

In Figure 2 there are 8 dashed, 8 dotted, and one dash-dot line between C1 and C2, for a total of 17 constraints. In general a set of n points describing a rigid body in three dimensions has six degrees of freedom (three for translation and three for rotation with respect to an external coordinate system); therefore, $3n - 6$ constraints are needed to make the body rigid. Each rigid polyhedron in Figure 2 contains 5 atoms and $(3 \times 5) - 6 = 9$ distance constraints (the C1-C2 constraint is shared by both). The ethane molecule as a whole can freely translate and rotate, and its dihedral angle gives it one internal degree of freedom; therefore, it should have $(3 \times 8) - 6 - 1 = 17$ constraints. This is exactly what Figure 2 shows.

Our method generalizes easily to molecules with more than one dihedral. Suppose there are two dihedrals dividing a molecule into three pieces.

The idea is to make each piece into a rigid polyhedron, then connect them together in pairs to allow rotation about the two dihedral axes. The end polyhedrons are treated just like the rigid methane groups in ethane. The middle polyhedron connects to two dihedral axes, but these dihedrals are necessarily distinct and do not interact with one another. Clearly we can extend this idea to any number of dihedrals, as long as they are truly free to rotate in the molecule (a single dihedral in a closed chain does not have full freedom, for instance).

Our procedure allows us to choose distance constraints for each rigid piece independently of the other pieces; that is, it is a strictly *local* procedure. Internal coordinate methods also divide the molecule into rigid pieces, but require a topology tree of interdependencies between the pieces [2,31]. The tree describes which pieces move when a particular dihedral rotates, and then a calculation is made involving all affected pieces to determine the constrained motion resulting from a dihedral rotation. The global interdependence of pieces results because internal coordinate methods are eliminating variables. We add constraints to the problem instead of eliminating variables, side-stepping the problem of calculating coupled rigid body motions.

4.1.1. Choosing the best constraints for a rigid piece.

We specified 9 distance constraints to make each of the five-atom polyhedrons in Figure 2 rigid. There are 10 possible atom-to-atom distance pairs, only 4 of which correspond to the length of a real chemical bond. This is not an unusual situation. In general, a set of n points has $(n^2 - n)/2$ possible distance pairs, of which we need $3n - 6$. A molecule with no rings has only $n - 1$ chemical bonds, so we face a growing surplus of constraint options as n increases.

For a rigid piece with $n > 4$ atoms we could add on distance constraints atom-by-atom as in [10]. This is a logical procedure, but it turns out that from an optimization perspective there is a "best" set of $3n - 6$ distance constraints. Each constraint is a quadratic equation in six unknown atom positions. For example, if R_{ij} is the fixed distance between atoms i and j, then the corresponding constraint equation is

$$(x_i - x_j)^2 + (y_i - y_j)^2 + (z_i - z_j)^2 = R_{ij}^2.$$

To solve the constrained optimization problem we will employ the transpose of the constraint Jacobian matrix defined by the gradients of the constraint equations. It is important to an optimization algorithm that this matrix be numerically well-conditioned; that is, that its columns be as linearly independent as possible. We define the "best" set of distance constraints as the set for which the matrix of gradients has the smallest condition number.

The best set of constraints is found automatically using a rank-revealing QR factorization [18]. We first assemble a matrix containing the gradient vectors for every possible distance constraint, arranged in any order. The rank-revealing factorization is a Gram-Schmidt orthogonalization procedure that chooses the next column (that is, constraint gradient) to be eliminated by examining the size of all remaining pivot elements. It passes through the entire matrix and returns an optimal ordering of constraints with their pivot sizes. The first $3n - 6$ constraints chosen by the factorization give a matrix with the desired small condition number. We use the LINPACK [13] subroutine dqrdc to perform the factorization.

A moment's consideration of Figure 2 reveals that we do not have total freedom in choosing our constraint set. We must make sure that we include the distance constraints that fix the lengths of any dihedral axes connected to our rigid piece (for ethane, this is the rigid bar C1-C2). Otherwise, two connecting rigid polyhedrons would have the freedom to shift along their common dihedral axis as well as rotate around it. Fortunately, dqrdc has the capability of forcing specified constraints to be in the front of the QR factorization matrix.

One other important detail needs discussion. What if the atoms forming a rigid section of the molecule are coplanar? In this case the rank-revealing QR factorization finds only $3n - 7$ nonzero pivots. The corresponding distance constraints force the atoms to be rigidly connected within a plane, but they do not force the atoms to remain coplanar (see [3, p. 95], for instance). As discussed in [40], subroutine dqrdc can easily detect nearly coplanar sets of atoms, and CCEMD [22] allows us to introduce fictitious noncoplanar atoms that keep the constraint Jacobian well-conditioned.

Let us summarize our method for choosing the distance constraints. A molecule is provided with bond lengths and bond angles already fixed at desired values, and a set of d dihedrals is specified.

Procedure for defining the distance constraints.
1. Using the d dihedral angles, partition the molecule into $d + 1$ nonoverlapping pieces, assigning each atom to exactly one piece. (We assume for simplicity that closed rings do not have free dihedrals.)
2. For the ith piece, i running from 1 to $d + 1$:

    a. Define a set S_i consisting of all the atoms in the piece, plus the atom on the opposite of every dihedral axis connected to this piece. Let n_i denote the number of atoms in S_i.

    b. Consider all pairwise distances between points in the set S_i, and construct a matrix whose columns are the gradients of these quadratic distance constraint equalities. The matrix has $3n_i$ rows and $(n_i^2 - n_i)/2$ columns.

    c. Perform a rank-revealing QR factorization on the matrix to find the best $3n_i - 6$ distance constraints. Force the factorization to include the distance constraints between atoms that define the dihedral axes touching this piece. (If the magnitude of pivot number $3n_i - 6$ is less than 10^{-2} times the magnitude of the first pivot, then add a fictitious noncoplanar atom to S_i and go back to step **2b**).

3. Combine all the distance constraints generated for each piece. Notice that the constraint between the two atoms defining a dihedral axis shows up twice because it was specifically included in each of the two pieces it joins. Keep just one of these two copies.

A straightforward calculation shows that this procedure generates the correct number of constraints. Let n be the total number of atoms in the molecule. The whole molecule has 6 external and d internal degrees of freedom, so this procedure should find a total of $3n - 6 - d$ distance constraints. Let d_i be the number of dihedral angles touching the ith piece. Since each dihedral touches exactly two different pieces, we see that

$$\sum_{i=1}^{d+1} d_i = 2d.$$

In step **2a** we included in S_i the atom on the opposite side of the axis for each dihedral touching a piece. Taking it back out gives a strict partitioning of the molecule, so

$$\sum_{i=1}^{d+1} (n_i - d_i) = n.$$

Now the total number of constraints collected in step **3** is just the sum of the number found for each piece minus the extraneous copy of each dihedral axis constraint. If we simplify this number using the previous equations, we obtain the correct number of constraints:

$$
\begin{aligned}
\sum_{i=1}^{d+1} (3n_i - 6) - d &= \sum_{i=1}^{d+1} (3n_i - 3d_i) + \sum_{i=1}^{d+1} (3d_i - 6) - d \\
&= 3n + 3(2d) - 6(d+1) - d \\
&= 3n - 6 - d.
\end{aligned}
$$

This procedure also works when free dihedrals are present in closed loops of atoms, provided we check that no duplicate distance constraints are generated between the rigid pieces comprising a loop. Duplicate constraints could conceivably arise between pairs of atoms on different dihedral axes within a loop. This is unlikely if the loop contains many dihedrals, as, for example, in our work with proteins where loops result from cysteine disulfide bridges.

4.2. An optimization algorithm for large-scale constrained optimization problems. In the previous section we showed how to define distance constraints so that only specific dihedral variables can change. This section describes the optimization algorithm used to solve the constrained energy minimization problem. The theory underlying this algorithm stems from the work of Byrd [6] and Omojokun [36] in the area of trust regions for equality constrained optimization. A general purpose software implementation of the algorithm called ETR was created with large-scale applications in mind [26,39].

We are faced with solving the constrained algebraic optimization problem

$$(4.1) \qquad \min \quad E(x_1, y_1, z_1, \ldots, x_n, y_n, z_n)$$

$$(4.2) \text{ subject to } \quad (x_i - x_j)^2 + (y_i - y_j)^2 + (z_i - z_j)^2 = R_{ij}^2, \quad \text{for } i, j \in \mathcal{D},$$

where the position of atom i in a Cartesian coordinate system is denoted by the triple (x_i, y_i, z_i), the fixed Euclidean distance between atoms i and j is the constant R_{ij}, and $\mathcal{D}$ is an index set containing the full list of distance constraints. The variables in this problem are the $3n$ coordinates of the atoms. The number of equality constraints is $3n - 6 - d$, which can be a large number if we specify only a few dihedrals to be free. The potential energy E is calculated in Cartesian coordinates by some molecular dynamics code in accordance with a given force field model. We assume that E is a continuously differentiable function of the variables and that its gradient (the negative of the force on every atom) can be calculated.

4.2.1. ETR algorithm for equality constrained optimization. The ETR algorithm is based on sequential quadratic programming (SQP), a standard approach for solving optimization problems with nonlinear equality constraints [14,15]. To use more general notation, let $x \in \mathbf{R}^N$ be the vector of variables, $f(x) : \mathbf{R}^N \to \mathbf{R}$ the function to be minimized, and $c(x) : \mathbf{R}^N \to \mathbf{R}^M$ the set of M equality constraints. The general constrained minimization problem is then written as

$$(4.3) \qquad \min_x f(x) \qquad \text{subject to} \quad c(x) = 0.$$

Basically, an SQP method adds a Lagrange multiplier variable λ_i for each of the M constraints and applies Newton's method to the resulting system of equations. The Newton method generates a sequence of iterates $\{x^0, x^1, \ldots, x^k, \ldots\}$ that converge to a solution of problem (4.3). For a given iterate x^k, a quadratic Taylor series expansion of (4.3) determines the SQP subproblem

$$(4.4) \qquad \min_p \quad f(x^k) + p^T \nabla f(x^k) + \frac{1}{2} p^T W(x^k, \lambda^k) p$$

$$(4.5) \qquad \text{subject to} \quad c_i(x^k) + p^T \nabla c_i(x^k) = 0, \quad \text{for } i = 1, \ldots, \text{M}.$$

We minimize this simpler subproblem to find the next iterate x^{k+1}. Note that $f(x^k)$, $c(x^k)$, $\nabla f(x^k)$, $\nabla c_i(x^k)$, and $W(x^k, \lambda^k)$ are constant quantities in the optimization subproblem. The variable $p \in \mathbf{R}^N$ is a distance vector from the point x^k, so that (4.4) represents a quadratic function in the components of p and (4.5) defines linear approximations to the constraints that p must satisfy. The matrix W is the Hessian of the Lagrangian

$$W(x^k, \lambda^k) = \nabla^2 f(x^k) + \sum_{i=1}^{M} \lambda_i^k \nabla^2 c_i(x^k),$$

which contains all second-order derivative information. The solution to (4.4)-(4.5) is a vector p^k, known as the *step* at the current iterate x^k. In a pure Newton method the next iterate is calculated directly as $x^{k+1} = p^k + x^k$.

For unconstrained optimization it is well-known that Newton's method only works if the starting iterate x^0 is already close to a solution. The same is true for constrained optimization. To make the SQP method work for an arbitrary starting guess we employ a *trust region* globalization technique. (Another common globalization technique uses line searches.) This is just a mechanism for judging the accuracy of subproblem (4.4)-(4.5) as a suitable model for the real problem (4.3), which is not quadratic. The trust region is a hypersphere about the point x^k with radius Δ. It is used to limit the length of the step p by appending to subproblem (4.4)-(4.5) the inequality

$$(4.6) \qquad\qquad \|p\|_2 \leq \Delta$$

($\| \cdot \|_2$ stands for the Euclidean norm of a vector). The idea is to make Δ smaller when the accuracy of the model seems poor. Since a significant amount of work goes into constructing and solving the SQP subproblem, however, a smart trust region algorithm acts to increase Δ when the accuracy seems good, thereby allowing bigger steps towards a solution.

There are a variety of ways to enforce the trust region inequality (4.6) while solving (4.4)-(4.5). We use the method of Byrd [6] and Omojokun [36] because it allows efficient solution of large-scale problems. Our software implementation of this method is called ETR (for *E*quality constrained optimization using *T*rust *R*egions) and is explained in detail in [26,39].

The ETR code computes a step p^k as the sum of two orthogonal vectors. One of these, the *vertical step*, attempts to satisfy the linearized constraint equations (4.5). If we collect the gradient vectors of each constraint into the N × M matrix

$$(4.7) \qquad A(x^k) = [\nabla c_1(x^k) \ \nabla c_2(x^k) \ \cdots \ \nabla c_M(x^k)],$$

then all M equations in (4.5) can be written collectively as $[A(x^k)]^T p^k + c(x^k) = 0$. It turns out that the vertical step $v \in \mathbf{R}^N$ computed by ETR always lies in the range space of $A(x^k)$; that is, v is a linear combination of the columns of $A(x^k)$.

The other part of p^k is called the *horizontal step*. It seeks to minimize the function (4.4) without disturbing the improvements made by v. To accomplish this it must be orthogonal to every constraint gradient, so we use sparse linear algebra techniques to construct an $N \times (N - M)$ matrix Z^k that satisfies the equation $[A(x^k)]^T Z^k = 0$. Then the horizontal step is expressed as a vector $Z^k u$, where $u \in \mathbf{R}^{N-M}$ are variables that are chosen to minimize (4.4) as much as possible. The $(N - M)$-dimensional subspace spanned by the columns of Z^k is the reduced subspace of problem (4.4)-(4.5); that is, the subspace left after imposing the M linearized constraints of (4.5). The vector u has one component corresponding to each degree of freedom in problem (4.4)-(4.5).

ETR forms the step as $p^k = v + Z^k u$. To judge the accuracy of this step we use the merit function $f(x^k + p^k) + \mu \| c(x^k + p^k) \|_2$, where $\mu > 0$ is a parameter that controls the relative importance of minimizing f and of satisfying the equality constraints. The method for choosing μ and other important details are documented in [26].

4.2.2. Applying the ETR algorithm. For a molecule with n atoms and d free dihedrals, the size of the optimization problem is $N = 3n$ and $M = 3n - 6 - d$. It takes one computation of the potential energy and interatomic forces in Cartesian coordinates to get $f(x^k)$ and $\nabla f(x^k)$. The values of the constraints and their gradients must also be computed to get $c(x^k)$ and $\nabla c_i(x^k)$; however, these are much cheaper (the computation is roughly equivalent to evaluating the energy and forces due to just the bond lengths). The second derivative information in W is usually approximated by a quasi-Newton matrix. A classical BFGS approximation is appropriate, but for large problems the storage requirements of the full matrix or its representation can become prohibitive. For this reason we use a compact limited memory BFGS [28,9] approximation for W, which stores only 10 vectors of length $3n$. Thus, the cost of setting up subproblem (4.4)-(4.5) is determined primarily by the cost of one evaluation of the potential energy and forces.

The vertical step v depends on $c(x^k)$ and the constraint gradients collected in $A(x^k)$, both of which were computed in setting up the SQP subproblem. ETR calculates v by treating the linearized constraint equations (4.5) as a linear least squares problem with a trust region; that is, by solving

$$\min_{v} \| c(x^k) + [A(x^k)]^T v \|_2 \qquad \text{subject to} \quad \| v \|_2 \leq 0.8\Delta.$$

Computing v involves some linear algebra operations with the matrix $A(x^k)$, but these are fairly cheap because we chose the distance constraints to make $A(x^k)$ well-conditioned, and because this is an extremely sparse matrix. As explained in [40], the sparsity of this matrix is a distinct advantage over methods which transform all Cartesian variables to dihedral variables.

Computation of the horizontal step is similar to a standard unconstrained quasi-Newton minimization of the potential energy. The main
differences are the presence of Lagrange multipliers in the quasi-Newton
Hessian approximation, and the restriction that the horizontal step be in
the form $Z^k u$. But the multipliers and Z^k both derive from A^k, which is
well-conditioned and computationally cheap to work with. Also, the complexity of dealing with Z^k is somewhat offset by the smaller size of the
reduced space minimization subproblem (its dimension is $N - M = 6 + d$).

In summary, we expect the cost of solving each SQP subproblem to be
dominated by the cost of evaluating the potential energy and interatomic
forces at the current iterate. There is one force evaluation per subproblem,
the same as in most unconstrained minimization algorithms. The extra
overhead of solving for nonlinear constraints is not large and should scale
linearly with the size of the molecule. If we assume that the potential
energy and interatomic forces can be calculated separately, then we obtain
the simple outline of the ETR algorithm shown below.

General description of the ETR algorithm for solving (4.1)-(4.2).

    1 Choose a molecular conformation and load initial atom positions
into x^0

    2 Make one energy evaluation to get $f(x^0)$ and $c(x^0)$

    3 Initialize $\Delta > 0$, $W^0 = I$, and $k = 0$

    4 Begin the main loop

        Make one force evaluation to get $\nabla f(x^k)$, and compute $A(x^k)$
using (4.7)

        Compute Z^k from $A(x^k)$

        Use $A(x^k)$ and $\nabla f(x^k)$ to compute Lagrange multiplier
estimates λ^k

  4a   **if** $\|\nabla f(x^k) - A(x^k)\lambda^k\|_\infty < \epsilon$ **and** $\|c(x^k)\|_\infty < \epsilon$ **then**
return *success*

  4b   Use $A(x^k)$ and $c(x^k)$ to compute a vertical step such that
$\|v\|_2 \leq 0.8\Delta$

        Use $\nabla f(x^k)$, W^k, and Z^k to compute a horizontal step with
$\|Z^k u\|_2^2 \leq \Delta^2 - \|v\|_2^2$

        Set $p^k = v + Z^k u$

        Make one energy evaluation at the new trial point to get
$f(x^k + p^k)$ and $c(x^k + p^k)$

        if the trial point is not a sufficiently good improvement
over x^k

            then $\Delta \leftarrow \gamma_1 \|p^k\|_2$, **goto 4b**

            else $x^{k+1} = x^k + p^k$, $\Delta \leftarrow \gamma_2 \Delta$, update the ℓ-BFGS
matrix W^k

    Increment k and **goto 4**

The main loop of the ETR algorithm sets up a series of SQP subproblems. The inner loop beginning at **4b** finds a suitable step p^k that solves the subproblem. ETR decides at **4a** that it has converged if first-order optimality conditions are satisfied to a tolerance ϵ. This means that every distance constraint is within ϵ of its proper length, and that every component of the reduced gradient is smaller than ϵ. The trust region size Δ is updated after every trial point using the parameters γ_1 and γ_2. For best algorithm performance we used $0.1 \leq \gamma_1 \leq 0.5$ and $2 \leq \gamma_2 \leq 5$, with the exact value determined by how much improvement was made in the merit function by the trial point (see [26] for details).

5. Computational test results. In this section we present computational results which show the relative effectiveness of two global search strategies. The GA and PDS algorithms were used to generate a large number of candidate starting conformations. The fitness of each candidate was evaluated by calculating a local energy minimum in torsion space using our constrained optimization method. From this data we plotted the low energy spectrum revealed by each global search scheme. We expect to observe that both the GA and PDS algorithms preferentially find low energy minima. Our experiments provide some quantification of the effectiveness of the search strategies. In addition, we will see whether the methods reveal the structure of the spectrum at higher energies.

5.1. Test problems. We chose two small synthetic peptides for this investigation, whose characteristics are summarized in Table 1. The peptides were prepared by using QUANTA [41] and were built with no hydrogens to reduce the CPU time for the energy calculations. The energy of each molecule was first gradient-minimized without constraints to form the reference conformation. Although the test molecules are fairly small, they possess a large number of distinct local energy minima. Assuming a simple three-fold symmetry about each dihedral, we expect on the order of $3^4 \approx 100$ distinct minima in torsion space for Thr-Ala, and $3^8 \approx 6500$ for Thr-Ala-Leu. Our objective is to identify all the minima within 10 kcal/mol of the global minimum.

TABLE 1
Test molecule characteristics

sequence	*number of atoms*	*number of dihedrals*
Thr-Ala	13	4
Thr-Ala-Leu	21	8

Each global search strategy varied the dihedral angles to generate a particular starting conformation of the molecule. Bond distances and bond angles were held fixed during this procedure. Our constrained optimization method was applied to each starting conformation and run until a local energy minimum in torsion space was found. The constraint equations were enforced so that rigid interatomic distances did not change by more than 10^{-5} Å. We carried out the local minimization to an unusually tight tolerance, requiring the Euclidean norm of the force vector expressed in dihedral variables to be less than 10^{-5} kcal/mol/Å. The tolerance is more accurate than the chemical model warrants, but our goal was to reliably distinguish between neighboring energy minima and provide a complete map of all low energy minima, similar to [43]. The size and average execution time of the constrained minimization problems are reported in Table 2. All calculations were performed on an SGI Power Challenge with a 75 MHz MIPS R8000 processor.

TABLE 2
Constrained optimization problem sizes

problem	number of unknowns	number of constraints	average CPU time
Thr-Ala	39	29	0.62 seconds
Thr-Ala-Leu	63	49	4.15 seconds

5.2. Energy minimization results. We accumulated data for three variants of the GA and PDS search strategies, which are listed in Table 3. The three runs differed primarily in the number of candidate conformations that were generated: 1,000 for the first run, 5,000 for the second, and 40,000 for the third run. In addition, we generated start points from a completely random distribution of dihedral angles. We do not suggest that this is a viable search strategy; it was used merely to help fill in the energy spectrum of the test molecules.

The three GA runs differed in the size of the population making up each generation and in the number of generations, as shown in Table 3. The third run also included niching operations between four sub-populations. The 'chromosomes' of the GA runs were 20-bit representations of the dihedral angle variables. We used a mutation rate of 0.01.

The three PDS runs were identical except for the total number of candidate points generated and the search scheme size. The search scheme size was set to 64 vertices for the Thr-Ala problem and to 256 vertices for the Thr-Ala-Leu problem. In addition, we modified the standard PDS algorithm so that it didn't generate any contraction points in the scheme. The effect of this modification is to allow the method to generate a coarser but broader scheme. Since we are using PDS solely as a search method and we are not concerned with finding a local minimum this allows us to explore more points overall.

The set of local energy minima found from each global search scheme was collected and analyzed for unique conformations. This was done by clustering together final conformations whose energies differed by less than 0.00005 kcal/mol and whose dihedral angles differed by less than 0.1 degree

TABLE 3

Description of global search strategies. Each line shows a search strategy for choosing dihedral angles to generate different molecular conformations.

0	Each dihedral treated as a uniformly random variable
1	GA with 20 generations, 50 individuals per generation
2	GA with 50 generations, 100 individuals per generation
3	GA with 100 generations, 100 individuals per generation, and 4 niches
4	PDS for a total of 1,000 conformations
5	PDS for a total of 5,000 conformations
6	PDS for a total of 40,000 conformations

rms. This first clustering criteria was applied to filter out "distinct" local minima which we feel were distinct only because the gradient minimization was not carried to a higher precision. Then a second clustering operation was applied to reduce the minima to a more chemically meaningful set. The members of each of these clusters had energies within 0.1 kcal/mol and dihedrals within 1.0 degree rms. To form these clusters, we examined the list of minima from lowest energy to highest and placed each conformation in an existing cluster if its energy and dihedrals differed by less than the tolerances from every other conformation already in the cluster.

We report the total number of each cluster type found by the different search strategies in Tables 4 and 5. Each line shows results for one of the global search strategies described in Table 3. The first two columns show the total number of starting points considered by each strategy (# *start pts*) and the lowest energy found. The next two pairs of columns each give the number of mathematically distinct local minima (# *math min*) and chemically relevant distinct minima (# *chem min*). The former are separated by at least 0.00005 kcal/mol or 0.1 degree rms and the latter by 0.1 kcal/mol or 1.0 degree rms. These tables also list the number of local minima found within 10 kcal/mol of the "global" minimum under the *Low energy minima* heading. We take as our estimate of the global minimum potential energy the smallest energy found by any method during our calculations.

Table 4 shows that the global minimum for the small Thr-Ala molecule was relatively easy to locate. However, the full set of chemically meaningful low energy states was harder to locate. From strategies 2 and 5 we see that searching over 5,000 candidate conformations turned up 80 % of the low energy minima. Strategies 3 and 6 show that up to 40,000 start points were needed to find all the low energy states.

The number of unknowns doubled in the molecule Thr-Ala-Leu and

TABLE 4

Search results for Thr-Ala. Low energy minima have energies < 17.9791 kcal/mol.

			All minima		Low energy minima	
strategy	*# start pts*	*lowest energy*	*# math min*	*# chem min*	*# math min*	*# chem min*
0	90,000	7.9791 kcal/mol	166	70	56	20
1	1,000	7.9791 kcal/mol	47	33	24	15
2	5,000	7.9791 kcal/mol	64	36	34	17
3	40,000	7.9791 kcal/mol	95	45	46	20
4	1,000	7.9791 kcal/mol	63	37	27	13
5	5,000	7.9791 kcal/mol	93	48	39	18
6	40,000	7.9791 kcal/mol	137	64	50	21

TABLE 5

Search results for Thr-Ala-Leu. Low energy minima have energies < −15.2766 kcal/mol.

			All minima		Low energy minima	
strategy	*# start pts*	*lowest energy*	*# math min*	*# chem min*	*# math min*	*# chem min*
0	90,000	−24.9003 kcal/mol	8270	4253	1530	695
1	1,000	−24.7243 kcal/mol	433	371	181	144
2	5,000	−25.2766 kcal/mol	1190	893	527	329
3	40,000	−25.2766 kcal/mol	3242	2042	1253	623
4	1,000	−24.7243 kcal/mol	281	238	98	78
5	5,000	−24.9920 kcal/mol	1495	1113	404	274
6	40,000	−25.2766 kcal/mol	5108	2964	1188	587

from Table 5 we see that both strategies were successful in finding the global minimum. The GA and PDS algorithms located approximately the same number of low energy minima, but PDS found significantly more high energy states for the same amount of work (compare strategy 2 with 5, and strategy 3 with 6).

We plotted the local energy minima found by each global search strategy in Figures 3 and 4. Each column in these figures shows the energy spectrum found by a particular search strategy. Each dash represents one local energy minimum after the second clustering operation was applied; that is, the figures correspond to data in the columns of Tables 4 and 5 headed by *# chem min*. In Figure 4, we have plotted all the energy minima, while Figure 5 shows only the minima within 10 kcal/mol of the apparent global minimum.

From Figure 3 we see that both the GA and PDS methods succeeded in mapping the full energy spectrum of the molecule. PDS found a slightly greater density of states among the higher energies for the same amount of work (compare columns 2 and 5 in the figure). The general structure of the spectrum was evident after only 1000 candidate conformations (columns 1 or 4).

Figure 4 plots only the low energy states of the spectrum of Thr-Ala-Leu. It is apparent that a large number of conformations must be examined to fill in the spectrum, especially at the lowest energies. With 5000 or fewer starting points (strategies 2 and 5), GA does a noticeably better job at finding the lower energy conformations than PDS. As the number of starting points increases however, this difference disappears.

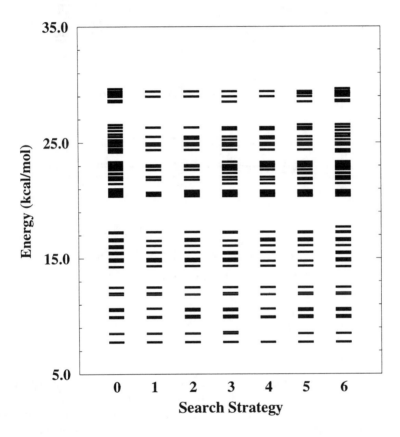

FIG. 3. *Energy spectrum for minimizing Thr-Ala in torsion space. Each mark shows the energy of a unique local minimum. The column numbers correspond to global search strategies.*

6. Summary. We have presented a comparison of two search methods, GA and PDS, for finding all of the local minima within a prescribed distance of the global minimum energy of a molecule. The GA method is an optimization algorithm designed to find the global minimum of a function. The PDS method is a local optimization method that we have employed as a search method. Although neither of the two methods was designed

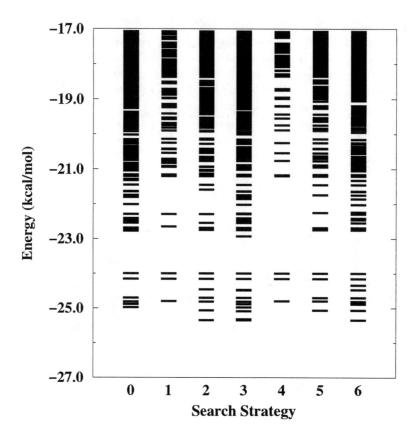

FIG. 4. *Low energy spectrum for minimizing Thr-Ala-Leu in torsion space. Each mark shows the energy of a unique local minimum. The column numbers correspond to global search strategies.*

for the purpose of finding more than one minimum, we have shown that in combination with a local gradient-based minimization method they can find a large number of local minima. Both methods tend to concentrate the computed minima towards the lower energies in the energy spectrum as the sample size of starting points is increased. In this sense, it can be argued that both methods would be appropriate for performing conformational searches.

We have also described some recent work [40] that uses distance constraints between atoms to allow potential energy minimization of molecules while holding all bond lengths and bond angles fixed. The constrained energy minima found by this method are identical to those found by minimizing in torsion space. Our method operates directly in Cartesian coordinates and avoids the usual difficulties associated with transforming to internal coordinates. We have presented a simple procedure for choosing appropriate distance constraints based on linear algebra considerations. It is simple because our constraints are determined solely by the atoms in a single rigid piece of the molecule – no analysis of coupled rigid body motions is needed. Our method requires the solution of a constrained optimization problem in $3n$ Cartesian unknowns instead of an unconstrained problem in d dihedral variables. By employing an optimization algorithm that exploits the sparsity structure of the constraints the new method has an added advantage over the apparently smaller minimization problem in internal coordinates.

REFERENCES

[1] R. ABAGYAN, M. TOTROV, AND D. KUZNETSOV, *ICM—a new method for protein modeling and design: applications to docking and structure prediction from the distorted native conformation*, J. Comp. Chem., 15:488–506, 1994.

[2] H. ABE, W. BRAUN, T. NOGUTI, AND N. GŌ, *Rapid calculation of first and second derivatives of conformational energy with respect to dihedral angles for proteins: general recurrent equations*, Comp. and Chem., 8:239–247, 1984.

[3] M.P. ALLEN AND D.J. TILDESLEY, *Computer Simulation of Liquids*, New York: Oxford UP, 1987.

[4] G. BOX AND K. WILSON, *On the experimental attainment of optimum conditions*, J. Royal Statistical Society, Series B, 13 (1951), pp. 1–45.

[5] B.R. BROOKS, R.E. BRUCCOLERI, B.D. OLAFSON, D.J. STATES, S. SWAMINATHA, AND M. KARPLUS, *CHARMM: a program for macromolecular energy, minimization, and dynamics calculations*, J. Comp. Chem., 4:187–217, 1983.

[6] R.H. BYRD, *Robust trust region methods for constrained optimization*, Third SIAM Conference on Optimization, Houston, 20 May 1987.

[7] R. BYRD, E. ESKOW, AND R. SCHNABEL, *A new large-scale global optimization method and its application to Lennard-Jones problems*, Tech. Report CU-CS-630-92, University of Colorado at Boulder, 1992.

[8] R. BYRD, E. ESKOW, R. SCHNABEL, AND S. SMITH, *Parallel global optimization: Numerical methods, dynamic scheduling methods, and applications to molecular configuration*, Tech. Report CU-CS-553-91, University of Colorado at Boulder, 1991.

[9] R.H. BYRD, J. NOCEDAL, R.B. SCHNABEL, *Representations of quasi-Newton matrices and their use in limited memory methods*, Math. Prog. (Ser. A), 63:129–156, 1994.

[10] G. CICCOTTI, M. FERRARIO, AND J.-P. RYCKAERT, *Molecular dynamics of rigid systems in cartesian coordinates: a general formulation*, Molec. Phys., 47:1253–1264, 1982.

[11] D. CVIJOVIC AND J. KLINOWSKI, *Taboo search: An approach to the multiple minima problem*, Science, 267 (1995), p. 664.

[12] J. DENNIS AND V. TORCZON, *Direct search methods on parallel machines*, SIAM J. Optimization, 1 (1991), pp. 448–474.

[13] J.J. DONGARRA, J.R. BUNCH, C.B. MOLER, AND G.W. STEWART, *LINPACK User's Guide*, Philadelphia: SIAM, 1979.

[14] R. FLETCHER, *Practical Methods of Optimization*, Second ed., Chichester, UK: Wiley & Sons, 1990.

[15] P. E. GILL, W. MURRAY, AND M. H. WRIGHT, *Practical Optimization*, London: Academic Press – Harcourt, 1981.

[16] N. GŌ AND H.A. SCHERAGA, *Ring closure and local conformational deformations of chain molecules*, Macromolecules, 3:178–187, 1970.

[17] D. GOLDBERG, *Genetic Algorithms in Search, Optimization, and Machine Learning*, Addison-Wesley, 1989.

[18] G. H. GOLUB AND C. F. VAN LOAN., *Matrix Computations*, Second ed., Baltimore: Johns Hopkins UP, 1991.

[19] M.R. HOARE, *Structure and dynamics of simple microclusters*, Adv. Chem. Phys., 40:49–135, 1979.

[20] R. HOOKE AND T. JEEVES, *Direct search solution of numerical and statistical problems*, J. Assoc. Comp. Mach., 8 (1961), pp. 212–229.

[21] A. HOWARD AND P. KOLLMAN, *An analysis of current methodologies for conformational searching of complex molecules*, J. Med. Chem., 31 (1988), pp. 1669–1675.

[22] R. JUDSON, D. BARSKY, T. FAULKNER, D. MCGARRAH, C. MELIUS, J. MEZA, E. MORI, AND T. PLANTENGA, *CCEMD - Center for Computational Engineering molecular dynamics: Theory and users' guide, version 2.2*, Tech. Report SAND95-8258, Sandia National Laboratories, 1995.

[23] R. JUDSON, M. COLVIN, J. MEZA, A. HUFFER, AND D. GUTIERREZ, *Do intelligent configuration search techniques outperform random search for large molecules?*, International Journal of Quantum Chemistry, 44 (1992), pp. 277–290.

[24] R. JUDSON, E. JAEGER, A. TREASURYWALA, AND M. PETERSON, *Conformational searching methods for small molecules II: A genetic algorithm approach*, J.Comp.Chem., 14 (1993), p. 1407.

[25] J. KOSTROWICKI, L. PIELA, B. CHERAYIL, AND H. SCHERAGA, *Performance of the diffusion equation method in searches for optimum structures of clusters of Lennard-Jones atoms*, J.Phys.Chem., 95 (1991), p. 4113.

[26] M. LALEE, J. NOCEDAL, AND T. PLANTENGA, *On the implementation of an algorithm for large-scale equality constrained optimization*, Submitted to SIAM J. Optimization, 1993.

[27] S.M. LE GRAND AND K.M. MERZ JR., *The application of the genetic algorithm to the minimization of potential energy functions*, Journal of Global Optimization, Vol. 3.1 (1993), pp. 49–66.

[28] D.C. LIU AND J. NOCEDAL, *On the limited memory BFGS method for large scale optimization*, Math. Prog. (Ser. B), 45:503–525, 1989.

[29] R. MAIER, J. ROSEN, AND G. XUE, *Discrete-continuous algorithm for molecular energy minimization*, Tech. Report 92-031, AHPCRC, 1992.

[30] C. MARANAS AND C. FLOUDAS, *A deterministic global optimization approach for molecular structure determination*, J.Chem.Phys., 100 (1994), p. 1247.

[31] A.K. MAZUR, V.E. DOROFEEV, AND R.A. ABAGYAN, *Derivation and testing of explicit equations of motion for polymers described by internal coordinates*, J. Comput. Phys., 92:261–272, 1991.

[32] J. MEZA, R. JUDSON, T. FAULKNER, AND A. TREASURYWALA, *A comparison of a direct search method and a genetic algorithm for conformational searching*, Tech. Report SAND95-8225, Sandia National Laboratories, 1995. To appear in the J. Comp. Chem., 1996.

[33] J. MEZA AND M. MARTINEZ, *Direct search methods for the molecular conformation problem*, Journal of Computational Chemistry, 15 (1994), pp. 627–632.

[34] F.A. MOMANY, R.F. MCGUIRE, A.W. BURGESS, AND H.A. SCHERAGA, *Geometric parameters, partial atomic charges, nonbonded interactions, hydrogen bond interactions, and intrinsic torsional potentials for the naturally occurring amino acids*, J. Phys. Chem., 79:2361–2381, 1975.

[35] J. NELDER AND R. MEAD, *A simplex method for function minimization*, Comput. J., 7 (1965), pp. 308–313.

[36] E.O. OMOJOKUN, *Trust region algorithms for optimization with nonlinear equality and inequality constraints*, Diss., Dept. of Computer Science, University of Colorado, 1989.

[37] P.M. PARDALOS, D. SHALLOWAY, AND G. XUE (EDITORS), *Optimization Methods for Computing Global Minima of Nonconvex Potential Energy Functions*, Journal of Global Optimization, Vol. 4.2 (1994), pp. 117–133.

[38] P.M. PARDALOS, D. SHALLOWAY, AND G. XUE (EDITORS), *Global Minimization of Nonconvex Energy Functions: Molecular Conformation and Protein Folding*, DIMACS Series, Vol. 23, American Mathematical Society (1996).

[39] T.D. PLANTENGA, *Large-scale nonlinear constrained optimization using trust regions*, Diss., Dept. of Electrical Engineering and Computer Science, Northwestern University, 1994.

[40] T. PLANTENGA AND R. JUDSON, *Energy minimization along dihedrals in cartesian coordinates using constrained optimization*, Tech. Report SAND95-8724, Sandia National Laboratories, 1995.

[41] QUANTA/CHARMM, Molecular Simulations, Inc. (Waltham MA, 1993). The results published were generated in part using the program QUANTA. This program was developed by Molecular Simulations, Inc.

[42] M. SAUNDERS, *Stochastic search for the conformations of icyclic hydrocarbons*, J.Comp.Chem., 10 (1989), p. 203.

[43] M. SAUNDERS, K. HOUK, Y.-D. WU, W. C. STILL, M. LIPTON, G. CHANG, AND W. C. GUIDA, *Conformations of cycloheptadecane. A comparison of methods for conformational searching*, J. Am. Chem. Soc., 112 (1990), pp. 1419–1427.

[44] S. SUNADA AND N. GŌ, *Small-amplitude protein conformational dynamics: second-order analytic relation between Cartesian coordinates and dihedral angles*, J. Comp. Chem., 16:328–336, 1995.

[45] S. WILSON AND W. CUI, *Applications of simulated annealing to peptides*, Biopolymers, 29 (1990), pp. 225–235.

[46] G. XUE, *Improvement on the Northby algorithm for molecular conformation: Better solutions*, Tech. Report 92-055, University of Minnesota, 1992.

ISSUES IN LARGE-SCALE GLOBAL MOLECULAR OPTIMIZATION

JORGE J. MORÉ* AND ZHIJUN WU*

Abstract. We discuss the formulation of optimization problems that arise in the study of distance geometry, ionic systems, and molecular clusters. We show that continuation techniques based on global smoothing are applicable to these molecular optimization problems, and we outline the issues that must be resolved in the solution of large-scale molecular optimization problems.

1. Introduction. We are concerned, in particular, with molecular optimization problems that arise in the study of protein structures in biophysical chemistry. If we adopt the hypothesis that the native protein structure corresponds to the global minimum of the protein energy [49,54], then the protein structure is determined by minimizing a potential energy function in conformational space. If the protein structure is determined from bounds on distances between pairs of atoms and other geometric constraints on the protein, then distance geometry [9,21] techniques are required. Both approaches require the solution of global optimization problems.

The problem of finding the least energy structure for a given molecular system arises not only in biological studies, but also in the study of large, confined ionic systems in plasma physics [20,44,50]. Configurations of large systems, for example, with 200,000 ions, are of special interest, because an important phase transition is expected by physicists for systems of this size. Such configurations require the solution of large-scale global minimization problems with up to $600,000$ variables – a computationally intensive problem, even for local minimization, because the problems are dense and each function evaluation requires order n^2 floating-point operations (flops).

Molecular optimization problems also arise in the study of clusters; for an introduction to the problems in this area, see the books edited by Reynolds [46] and Haberland [19]. Much of the interest in clusters is due to unexpected theoretical and practical results, such as the discovery of the stable carbon cluster C_{60}. Theoretical properties of clusters usually are determined by molecular dynamics simulation or by potential energy minimization. Small argon clusters have received considerable attention in the past (see, for example, the classical studies of Hoare [24] and Northby [40]), since they have simple potentials and structure, but current interest centers on clusters with more involved potentials.

The molecular optimization problems that we have mentioned are dif-

* ARGONNE NATIONAL LABORATORY, 9700 South Cass Avenue, Argonne, Illinois 60439. This work was supported by the Mathematical, Information, and Computational Sciences Division subprogram of the Office of Computational and Technology Research, U.S. Department of Energy, under Contract W-31-109-Eng-38, and by the Argonne Director's Individual Investigator Program.

ficult because the presence of a large number of local minimizers, even for systems with a small number of atoms, creates numerous regions of attraction for local searches. We have been using global smoothing and continuation techniques for these problems. In this approach the Gaussian transform is used to map the objective function into a smoother function with fewer local minimizers, and an optimization procedure is applied to the transformed function, tracing the minimizers as the function is gradually changed back to the original function. A transformed function is a coarse approximation to the original function, with small and narrow minimizers being removed while the overall structure is maintained. This property allows the optimization procedure to skip less interesting local minimizers and to concentrate on regions with average low function values, where a global minimizer is most likely to be located.

The Gaussian transform has been used in many areas, in particular, stochastic and non-smooth optimization. For a sampling of this work, see Katkovnik and Kulchitskii [28], Rubinstein [47], Kreimer and Rubinstein [32], and Ermoliev, Norkin, and Wets [10]. We also note that Stillinger [56] used the Gaussian transform to study the scaling behavior of many-body potentials.

The Gaussian transform is a central component in the diffusion equation method for protein conformation, proposed by Scheraga and coworkers [43,29,30,31,49] for global molecular optimization. A similar approach was used in the packet annealing algorithm of Shalloway [54,53] and in the algorithms of Coleman, Shalloway, and Wu [7,8] for molecular conformation problems. Recent work on the Gaussian transform from a mathematical and computational point of view can be found in Wu [60] and Moré and Wu [38,37]. Anisotropic smoothing, a generalization of the Gaussian smoothing, has also been considered by many of the workers in this area. For recent work in this area, see Orešič and Shalloway [41], Straub, Ma, and Arena [58], and Wu [60]. Moré and Wu [39] have also extended the smoothing properties of the Gaussian transform to transformations with other density functions. For a recent review of the smoothing approach for global molecular optimization, see Pardalos, Shalloway, and Xue [42] and Straub [57].

While Gaussian smoothing is a promising approach for the solution of molecular optimization problems, many theoretical and computational issues need additional attention. In this paper we review current work on distance geometry problems, and indicate how these results can be extended to ionic systems and molecular clusters. We focus on issues that must be addressed in large-scale molecular optimization problems.

In Section 2 we discuss the formulation of molecular optimization problems that arise in distance geometry calculations, ionic systems, and molecular clusters. Section 3 is a review of the properties of the Gaussian transform. In particular, we outline the techniques used to compute the Gaussian transform for molecular optimization problems. In Section 4 we discuss computational experiments carried out with a simple continuation algorithm, and we show how problem formulation affects the choice of optimization procedure in the continuation algorithm. We conclude the paper in Section 5 with a critical review of issues that must be addressed in order to solve large-scale molecular optimization problems on high-performance architectures.

2. Molecular optimization problems. A typical molecular optimization problem is to determine a structure with minimal potential energy. In some cases, the position of the atoms in the structure must also satisfy certain physical constraints. In this section we review three molecular optimization problems and related work.

2.1. Distance geometry. Distance geometry problems arise in the interpretation of nuclear magnetic resonance (NMR) data and in the determination of protein structures. For a general review of the distance geometry problem and its relationship to macromolecular modeling, see Crippen and Havel [9], Havel [21], Kuntz, Thomason, and Oshiro [33], and Brünger and Nilges [4].

A distance geometry problem is specified by a subset S of all atom pairs and by the distances $\delta_{i,j}$ between atoms i and j for $(i,j) \in S$. A solution to the distance geometry problem is a set of positions $x_1, \ldots, x_m$ in $\mathbb{R}^3$ such that

$$(2.1) \qquad \|x_i - x_j\| = \delta_{i,j}, \qquad (i,j) \in S.$$

Usually, S is sparse; in other words, only a small subset of distances is known.

In practice, lower and upper bounds on the distances are specified instead of their exact values. The distance geometry problem with lower and upper bounds is to find positions $x_1, \ldots, x_m$ such that

$$(2.2) \qquad l_{i,j} \le \|x_i - x_j\| \le u_{i,j}, \qquad (i,j) \in S,$$

where $l_{i,j}$ and $u_{i,j}$ are lower and upper bounds on the distance constraints, respectively. An important case of this problem is to obtain an ε-optimal solution to the distance geometry problem, that is, positions $x_1, \ldots, x_m$ such that

$$(2.3) \qquad \left| \|x_i - x_j\| - \delta_{i,j} \right| \le \varepsilon, \qquad (i,j) \in S$$

for some $\varepsilon > 0$. An ε-optimal solution is useful when the exact solution to the problem (2.1) does not exist because of small errors in the data. This situation can happen, for example, when the triangle inequality

$$\delta_{i,j} \le \delta_{i,k} + \delta_{k,j}$$

is violated for atoms $\{i, j, k\}$ because of possible inconsistencies in the experimental data.

The distance geometry problem (2.1) is computationally intractable because the restriction of the distance geometry problem to a one-dimensional space is equivalent to the set partition problem, which is known to be NP-complete [11]. Saxe [48] shows that k-dimensional distance geometry problems are strongly NP-hard for all $k \geq 1$. The following result of Moré and Wu [37] shows that obtaining an approximate solution to the distance geometry problem is also NP-hard.

THEOREM 2.1. *Determining an ε-optimal solution to the distance geometry problem in $\mathbb{R}$ is NP-hard.*

The distance geometry problems that we have described can be formulated as global optimization problems for which the constraints are satisfied at a global minimizer of the problem. A simple formulation is in terms of finding the global minimum of the function

$$(2.4) \qquad f(x) = \sum_{i,j \in \mathcal{S}} p_{i,j}(x_i - x_j),$$

where the pairwise potential $p_{i,j} : \mathbb{R}^n \mapsto \mathbb{R}$ is defined for problem (2.1) by

$$(2.5) \qquad p_{i,j}(x) = \left(\|x\|^2 - \delta_{i,j}^2 \right)^2,$$

while Crippen and Havel [9] suggested that for problem (2.2)

$$(2.6) \quad p_{i,j}(x) = \min^2 \left\{ \frac{\|x\|^2 - l_{i,j}^2}{l_{i,j}^2}, 0 \right\} + \max^2 \left\{ \frac{\|x\|^2 - u_{i,j}^2}{u_{i,j}^2}, 0 \right\}.$$

Clearly, $x = \{x_1, \dots, x_m\}$ solves the distance geometry problem if and only if x is a global minimizer of f and $f(x) = 0$.

Special optimization algorithms have been developed for solving the distance geometry problem (2.1). For example, Hendrickson [22,23] used a graph-theoretic viewpoint to develop algorithms that test the uniqueness and rigidity of the distance graph. These algorithms can be used to reduce the problem into smaller, easier subproblems. Glunt, Hayden, and Raydan [13,14] have proposed a special gradient method for determining a local minimizer of the problem defined by (2.4) with

$$p_{i,j}(x) = \left(\|x\| - \delta_{i,j} \right)^2.$$

Al-Homidan and Fletcher [1] have done related work on a hybrid algorithm that combines an alternating projection method with a quasi-Newton method.

If all pairwise distances are known and a solution exists, then the solution of the distance geometry problem (2.1) can be determined (Blumenthal [3, Section 43], Crippen and Havel [9, Section 6.3]) by computing the largest three eigenvalues and eigenvectors of the rank-3 positive semidefinite matrix $A \in \mathbb{R}^{m \times m}$ defined by

$$a_{i,j} = \tfrac{1}{2} \left(\delta_i^2 + \delta_j^2 - \delta_{i,j}^2 \right), \qquad i, j = 1, \ldots, m,$$

where $\delta_i = \|x_i - x_0\|$, and $x_0 \in \mathbb{R}^n$ is a convex combination of $x_1, \ldots, x_m$. Note that if x_0 is a convex combination of $x_1, \ldots, x_m$, then δ_i can be expressed in terms of $\delta_{i,j}$. In practice, x_0 is the centroid of $x_1, \ldots, x_m$. We can determine the coordinates $x_1, \ldots, x_m$ by noting that the identity

$$2(x_i - x_0)^T (x_j - x_0) = \|x_i - x_0\|^2 + \|x_j - x_0\|^2 - \|x_i - x_j\|^2$$

implies that $A = B^T B$ is a positive semidefinite rank-3 matrix with

$$B = (x_1 - x_0, \ldots, x_m - x_0).$$

Hence, the vectors $x_1, \ldots, x_m$ can be determined by computing the largest three eigenvalues and eigenvectors of the rank-3 matrix A. Alternatively, we could use the Cholesky decomposition with diagonal pivoting.

These results show that the distance geometry problem can be solved in polynomial time if all distances are given. This does not contradict the result that the distance geometry problem is NP-hard, because we are restricting the general problem by requiring that all distances be given.

In practice, only a small subset of the distances is known, and there are experimental errors in the data, so the above procedure cannot be used. However, an extension of this procedure is employed by the **embed** algorithm (see Crippen and Havel [9], and Havel [21]) in practical distance geometry calculations. In the first phase of the **embed** algorithm, the sparse set of distance constraints is extended by using the relationships

$$u_{i,j} = \min (u_{i,j}, u_{i,k} + u_{k,j}), \qquad l_{i,j} = \max (l_{i,j}, l_{i,k} - u_{k,j}, l_{j,k} - u_{k,i}).$$

Given a full set of bounds, distances $\delta_{i,j} \in [l_{i,j}, u_{i,j}]$ are chosen, and an attempt is made to compute coordinates $x_1, \ldots, x_m$ as in the above procedure. This attempt usually fails, but it can be used to generate a rank-3 approximation to A, which leads to an approximation to the solution of problem (2.1). This approximation can be refined by minimizing a function of the form (2.4,2.5).

The **embed** algorithm, as described above, may require many trial choices of $\delta_{i,j}$ in $[l_{i,j}, u_{i,j}]$ before a solution to problem (2.2) is found. Current implementations of the **embed** algorithm use a local minimizer of the problem defined by (2.4) and (2.5) as a starting point for a simulated annealing procedure. In Section 4 we will outline the proposal of Moré and Wu [37] for finding a solution of the distance geometry problem (2.2) by directly determining a global minimizer of the function defined by (2.4) and (2.6).

2.2. Ionic systems. The potential energy for a confined ionic system of m ions located at $x_1, \ldots, x_m$ in R^3 can be modeled, for example, by a function of the form

$$(2.7) \qquad f(x) = \sum_{i \neq j} p_{i,j}(x_i - x_j) + \sum_{i=1}^{m} \|x_i\|^2,$$

where $p_{i,j} : \mathbb{R}^3 \mapsto \mathbb{R}$ is defined by

$$(2.8) \qquad p_{i,j}(x) = v(\|x\|), \qquad v(r) = r^{-1}.$$

Hasse and Schiffer [20], Rafac, Schiffer, Hangst, Dubin, and Wales [44], and Schiffer [50] studied configurations of confined ionic systems via molecular dynamics simulation. The results for small systems showed that ionic systems have a layered shell structures, with the number of shells increasing as the number of atoms in the system increases. The distribution of the ions over the shells also varies with different systems. For example, Figure 2.1 shows that the system of 60 ions has two shells with 12 ions in the inner shell and 48 ions in the outer shell, while the system of 61 ions has three shells with a single ion (at the center of the system) as the innermost shell.

FIG. 2.1. *Confined ionic system with 60 ions (left) and 61 atoms (right).*

Experiments indicate that as the number of atoms increases, the boundaries between the shells become blurred, and eventually the system achieves a crystal form. Therefore, there must be a phase transition from a system of layered shells to a body-centered cubic lattice, a standard crystal structure. This phase transition is of special physical interest, but in order to locate the transition, configurations for very large systems must be determined. The nature of the transition and the number of atoms at which the transition occurs is under investigation, but current work has focused on systems with $200,000$ ions. This calculation would be prohibitively expensive for a molecular dynamics simulation, but may be possible by minimizing the potential energy function. We have found in our preliminary studies that configurations for most systems with up to 100 ions can be determined by a single local minimization. Of course, for large systems, global optimization algorithms are required to obtain the least energy configurations, which represent the most stable structures of the systems.

2.3. Molecular clusters. A cluster is a group of identical molecules with specific geometrical and chemical properties. Clusters of chemical importance include, for example, argon and carbon clusters. A fundamental problem in cluster science is to determine the geometrical structure of clusters in their lowest energy states. Related problems include structure changes from clusters to bulk matter and low energy paths between stable states. For a general review of these topics, see Haberland [19].

Clusters of argon molecules were first studied by Hoare and coworkers [26,25,24]. Northby [40] obtained the structures for clusters with up to 147 molecules using a lattice search algorithm, which later was improved and used for even larger clusters by Xue [61]. Results for small argon clusters have also been obtained by general-purpose algorithms such as the diffusion equation method [30], the packet annealing algorithm [54], the stochastic search method [6,5], and the effective energy simulated annealing algorithm [7].

Argon clusters have been heavily studied because the potentials and structure of these clusters are relatively simple. Argon clusters usually are modeled by the Lennard-Jones potential

$$f(x) = \sum_{i \neq j} p_{i,j}(x_i - x_j),$$

where $p_{i,j} : \mathbb{R}^3 \mapsto \mathbb{R}$ is defined by

$$p_{i,j}(x) = v(\|x\|), \qquad v(r) = r^{-12} - 2r^{-6},$$

or by the Morse potential

$$v(r) = (1 - \exp[\alpha(1 - r)])^2 - 1,$$

for some positive constant α. For example, Hoare [24] used $\alpha = 3$.

Potentials for other clusters can be more involved. For example, in the study of metal clusters [46,19,27] it is common to use potentials of the form

$$f(x) = \sum_{i \neq j} p_{i,j}(x_i - x_j) - \sum_{j=1}^{n} \left(\sum_{i \neq j} q_{i,j}(x_i - x_j) \right)^{1/2},$$

where $p_{i,j} : \mathbb{R}^3 \mapsto \mathbb{R}$ and $q_{i,j} : \mathbb{R}^3 \mapsto \mathbb{R}$ are of the form $v(\|x\|)$ with

$$v(r) = \alpha \exp\left[-\beta(r - 1)\right],$$

for positive constants α and β. Note that these potentials are functions of the pairwise distance between atoms, and that they decay rapidly as r approaches infinity. The potential for ionic systems, on the other hand, decays slowly as r approaches infinity.

Cluster problems are difficult for most global optimization strategies because they tend to have a large number of local minimizers that act as points of attraction for any local minimizer. For argon clusters, Hoare [24] found that systems with $6 \leq m \leq 13$ atoms had

$$2, \ 4, \ 8, \ 18, \ 57, \ 145, \ 366, \ 989$$

different minima, respectively, and on the basis of this observation conjectured that the number of minima grew like $\exp(m^2)$. We are usually interested in global minimizers, but local minimizers with low function values are also of interest because they represent the most stable structures. For a discussion of these issues see, for example, Jellinek [27].

3. Smoothing transformations. The global continuation approach to finding the global minimizer is to transform the function into a smoother function with fewer local minimizers, apply an optimization algorithm to the transformed function, and trace the minimizers back to the original function. This approach is well suited for problems with many local minimizers. As already noted, molecular optimization problems tend to have a large number of local minimizers.

A transformed function is a coarse approximation to the original function, with small and narrow minimizers being removed, while the overall structure of the function is maintained. This property allows the optimization algorithm to skip less interesting local minimizers and to concentrate on regions with average low function values, where a global minimizer is most likely to be located.

The smoothing transform, called the Gaussian transform, depends on a parameter λ that controls the degree of smoothing. The original function is obtained if $\lambda = 0$, while smoother functions are obtained as λ increases.

DEFINITION 3.1. *The Gaussian transform* $\langle f \rangle_\lambda$ *of a function* $f : \mathbb{R}^n \mapsto \mathbb{R}$ *is*

$$(3.1) \qquad \langle f \rangle_\lambda(x) = \frac{1}{\pi^{n/2} \lambda^n} \int_{\mathbb{R}^n} f(y) \exp\left(-\frac{\|y - x\|^2}{\lambda^2}\right) dy.$$

The value $\langle f \rangle_\lambda(x)$ is an average of f in a neighborhood of x, with the relative size of this neighborhood controlled by the parameter λ. The size of the neighborhood decreases as λ decreases so that when $\lambda = 0$, the neighborhood is the center x. The Gaussian transform $\langle f \rangle_\lambda$ can also be viewed as the expected value of f with respect to the Gaussian density function

$$\rho_\lambda(y) = \frac{1}{\pi^{n/2}\lambda^n} \exp\left(-\frac{\|y\|^2}{\lambda^2}\right).$$

For the mathematical properties of the Gaussian transform, readers are referred to Wu [60] and Moré and Wu [38]. Note that a generalization of the Gaussian transform, the anisotropic Gaussian transform, can be defined by replacing λ with a matrix. We discuss this generalization in Section 5.2. Moré and Wu [39] show that the Gaussian transform can also be generalized by using other density functions.

Motivation for the Gaussian transform can be obtained by showing that the Gaussian transform of the two-dimensional version of the Griewank function

$$(3.2) \qquad f(x) = 1 + \sum_{i=1}^{n}\left(\frac{x_i^2}{200}\right) - \prod_{i=1}^{n}\cos\left(\frac{x_i}{\sqrt{i}}\right)$$

removes local minimizers. This function was constructed by Griewank [17] to test global optimization algorithms on problems with a large number of local minimizers. Figure 3.1 shows plots of the Griewank function and the Gaussian transform

$$(3.3) \qquad \langle f \rangle_\lambda(x) = 1 + \sum_{i=1}^{n}\left(\frac{x_i^2}{200} + \frac{\lambda^2}{400}\right) - \prod_{i=1}^{n}\exp\left(-\frac{\lambda^2}{4i}\right)\cos\left(\frac{x_i}{\sqrt{i}}\right)$$

of the Griewank function; justification for (3.3) as the Gaussian transform of the Griewank function will be provided shortly.

Figure 3.1 shows that the Gaussian transform reduces the number of minimizers as λ increases, and that the global minimizer of the original function can be found by applying a local minimization algorithm to the transformed functions and tracing the minimizers back to the original function. Although these plots are suggestive, it is important to keep in mind that the global smoothing approach is not guaranteed to succeed in all cases.

The following result of Wu [60] explains why the Gaussian transform reduces the high-frequency components of the function and eliminates local minimizers as λ increases.

THEOREM 3.2. *If $\widehat{f} : \mathbb{R}^n \mapsto \mathbb{C}$ is the Fourier transform of $f : \mathbb{R}^n \mapsto \mathbb{R}$, then*

$$\left|\widehat{\langle f \rangle_\lambda}(\omega)\right| = \exp\left(-\tfrac{1}{4}\lambda^2\|w\|^2\right)\left|\widehat{f}(\omega)\right|.$$

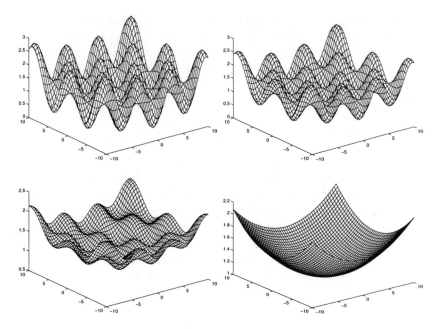

FIG. 3.1. *The Gaussian transform for the Griewank function. The original function* ($\lambda = 0$) *is in the top left corner, with* $\lambda = 1$ *in the top right corner,* $\lambda = 2$ *in the bottom left corner, and* $\lambda = 4$ *in the bottom right corner.*

Theorem 3.2 shows that $\widehat{\langle f \rangle}_\lambda(\omega)$, the component of $\langle f \rangle_\lambda$ for frequency ω, can be made arbitrarily small if $\lambda \|\omega\|$ is sufficiently large. In particular, the high-frequency components are removed when λ is large.

Computing the Gaussian transform usually requires the evaluation of n-dimensional integrals, but for many functions that arise in practice, it is possible to compute the Gaussian transform explicitly in terms of one-dimensional transforms. In particular, if the function is *decomposable*, that is, if the function $f : \mathbb{R}^n \mapsto \mathbb{R}$ can be written in the form

$$f(x) = \sum_{k=1}^{m} \left(\prod_{j=1}^{n} f_{k,j}(x_j) \right),$$

for some set of functions $\{f_{k,j}\}$, where $f_{k,j} : \mathbb{R} \mapsto \mathbb{R}$, then a simple computation shows that

$$\langle f \rangle_\lambda(x) = \sum_{k=1}^{m} \left(\prod_{j=1}^{n} \langle f_{k,j} \rangle_\lambda(x_j) \right).$$

Thus, computing $\langle f \rangle_\lambda$ for a decomposable function requires the computation of only one-dimensional transforms $\langle f_{k,j} \rangle_\lambda$.

The Gaussian transform of polynomials and other analytic functions have been determined by Kostrowicki and Piela [29] by using an alternative definition of the Gaussian transform. For a derivation that uses definition (3.1), see Moré and Wu [38]. These results show, in particular, that

$$t \mapsto t^2 + \tfrac{1}{2}\lambda^2, \qquad t \mapsto \cos(\alpha t)\exp\left(-\tfrac{1}{4}\alpha^2\lambda^2\right),$$

are the Gaussian transforms of the functions $t \mapsto t^2$ and $t \mapsto \cos(\alpha t)$, respectively. Since (3.2) is decomposable, the Gaussian transform of the Griewank function is given by (3.3).

For distance geometry problems, as well as for many other molecular problems, we are interested in transforming a class of functions defined in terms of the distances between pairs of atoms. Given functions $p_{i,j} : \mathbb{R}^p \mapsto \mathbb{R}$ of the distances between atoms i and j, we consider a general function

$$(3.4) \qquad f(x) = \sum_{i,j \in \mathcal{S}} p_{i,j}(x_i - x_j),$$

where $\mathcal{S}$ is some subset of all pairs of atoms, and $x_i \in \mathbb{R}^p$ is the position of the i-th atom. We are concerned with three-dimensional problems where $p = 3$.

The following result of Moré and Wu [38] shows that computing the Gaussian transform of (3.4) requires only the Gaussian transform of $p_{i,j}$.

THEOREM 3.3. *If $f : \mathbb{R}^n \mapsto \mathbb{R}$ and $h : \mathbb{R}^p \mapsto \mathbb{R}$ are related by*

$$f(x) = h(P^T x),$$

for some matrix $P \in \mathbb{R}^{n \times p}$ such that $P^T P = \sigma^2 I$, then

$$\langle f \rangle_\lambda(x) = \langle h \rangle_{\sigma\lambda}(P^T x).$$

Theorem 3.3 reduces the computation of the Gaussian transform of f, which is defined on $\mathbb{R}^n$, to the computation of the Gaussian transform of h, which is defined on $\mathbb{R}^p$. As an application of this result, note that

$$\langle f \rangle_\lambda(x) = \sum_{i,j \in \mathcal{S}} \langle p_{i,j} \rangle_{\sqrt{2}\lambda}(x_i - x_j)$$

is the Gaussian transform for the function defined by (3.4). In this case f is defined on $\mathbb{R}^{pn}$, but $p_{i,j}$ is defined on $\mathbb{R}^p$.

In some applications we can compute the Gaussian transform $\langle p_{i,j} \rangle_\lambda$ explicitly. For example, in the distance geometry problem (2.1), the function $p_{i,j} : \mathbb{R}^p \mapsto \mathbb{R}$ is defined by

$$(3.5) \qquad p_{i,j}(x) = \left(\|x\|^2 - \delta_{i,j}^2 \right)^2.$$

This function is decomposable. Moreover, the Gaussian transform is explicitly given by

$$(3.6) \quad \langle p_{i,j} \rangle_\lambda(x) = p_{i,j}(x) + [3 + (p-1)]\lambda^2 \|x\|^2 + \tfrac{1}{4}p(p+2)\lambda^4 - p\delta_{i,j}^2\lambda^2.$$

For a derivation of this result, see Moré and Wu [38].

As we have noted in Section 2, most of the potentials used in molecular optimization problems are sums of functions of the form $f(x) = h(\|x\|)$ for some mapping $h : \mathbb{R} \mapsto \mathbb{R}$. The following result of Moré and Wu [37] shows that the Gaussian transform of f can be expressed in terms of one-dimensional integrals.

THEOREM 3.4. *If $f : \mathbb{R}^3 \mapsto \mathbb{R}$ is of the form $f(x) = h(\|x\|)$ for some function $h : \mathbb{R} \mapsto \mathbb{R}$, then*

$$\langle f \rangle_\lambda(x) = \frac{1}{\lambda\sqrt{\pi}r} \int_0^\infty sh(s) \left[\exp\left(-\frac{(r-s)^2}{\lambda^2}\right) - \exp\left(-\frac{(r+s)^2}{\lambda^2}\right) \right] ds,$$

where $r = \|x\|$. If the mapping h is an even function, then

$$\langle f \rangle_\lambda(x) = \frac{1}{\lambda\sqrt{\pi}r} \int_{-\infty}^{+\infty} sh(s) \exp\left(-\frac{(r-s)^2}{\lambda^2}\right) ds.$$

Theorem 3.4 can be used to compute the Gaussian transform for molecular optimization problems. As an example, note that the potential function (2.7) used for the confined ionic system has $h(r) = r^{-1}$, and thus the Gaussian transform is

$$\begin{aligned}
\langle h \rangle_\lambda(x) &= \frac{1}{\lambda\sqrt{\pi}r} \int_0^\infty \left[\exp\left(-\frac{(r-s)^2}{\lambda^2}\right) - \exp\left(-\frac{(r+s)^2}{\lambda^2}\right) \right] ds \\
&= \frac{1}{\sqrt{\pi}r} \left[\int_{-\infty}^{r/\lambda} \exp(-t^2)dt - \int_{r/\lambda}^{+\infty} \exp(-t^2)dt \right] \\
&= \frac{1}{\sqrt{\pi}r} \int_{-r/\lambda}^{+r/\lambda} \exp(-t^2)dt = \frac{2}{\sqrt{\pi}r}\mathrm{erf}(r/\lambda),
\end{aligned}$$

where erf is the standard error function.

Theorem 3.4 reduces the computation of the Gaussian transform to the computation of one-dimensional integrals. We can compute these integrals with standard techniques for numerical integration (for example, an adaptive quadrature), but these techniques usually require a large number of function evaluations. An alternative approach is to use a Gauss-Hermite approximation, as suggested by Moré and Wu [37]. The Gauss-Hermite approximation to the one-dimensional transform

$$\langle f \rangle_\lambda(x) = \frac{1}{\sqrt{\pi}} \int_{-\infty}^{+\infty} f(x + \lambda s) \exp(-s^2) \, ds$$

is obtained by using Gaussian quadratures on the integral. Hence,

$$(3.7) \qquad \langle f \rangle_{\lambda,q}(x) = \frac{1}{\sqrt{\pi}} \sum_{i=1}^q w_i f(x + \lambda s_i)$$

is the Gauss-Hermite transform, where w_i and s_i are, respectively, the standard weights and nodes for Gaussian quadratures. The weights and nodes can be found in the tables of Stroud and Secrest [59] or computed with the **gauss** subroutine in **ORTHOPOL** [12] if the function h is even. For general h we should use a Gaussian quadrature for the semi-infinite interval $[0, +\infty)$, as advocated by Steen, Byrne, and Gelbard [55].

The Gauss-Hermite transform (3.7) can be viewed as a discrete transform, with carefully chosen sample points. The computation of the Gauss-Hermite transform requires q function evaluations, but even for modest values of q we obtain a good approximation to the Gaussian transform.

THEOREM 3.5. *Let* $\langle f \rangle_{\lambda, q}$ *be the transformation of* $f : \mathbb{R} \mapsto \mathbb{R}$ *defined in (3.7). If* $f^{(l)}$ *is piecewise continuous on* $\mathbb{R}$ *for some* $l \leq 2q$, *there is a constant* μ_l, *independent of* f, *such that*

$$|\langle f \rangle_\lambda(x) - \langle f \rangle_{\lambda, q}(x)| \leq \mu_l \lambda^l \sigma(x),$$

where

$$\sigma(x) = \left\{ \int_{-\infty}^{+\infty} \exp(-s^2) \left| f^{(l)}(x + \lambda s) \right|^2 ds \right\}^{1/2}.$$

Theorem 3.5 is due to Moré and Wu [37]. This result shows that (3.7) is a good approximation to $\langle f \rangle_\lambda$ provided $\lambda < 1$, but that the accuracy is likely to deteriorate if $\lambda > 1$. This is not a serious difficulty because for large λ we use (3.7) only to guide an algorithm to a global minimizer, but for small λ we work with the original function f.

4. Computational experiments. Given the Gaussian transform $\langle f \rangle_\lambda$, we can use a continuation algorithm to trace a minimizer of $\langle f \rangle_\lambda$. In this section we provide an overview of computational experiments carried out with a simple continuation algorithm that uses a sequence of continuation parameters

$$\lambda_0 > \lambda_1 > \cdots > \lambda_p = 0.$$

An optimization algorithm is used to determine a minimizer x_{k+1} of $\langle f \rangle_{\lambda_k}$. For λ_0 we can use any starting point, but for λ_k with $k > 0$, it is reasonable to use x_k as the starting point. Algorithm gmin provides an outline of our continuation algorithm:

Algorithm gmin
 Choose a random vector $x_0 \in \mathbb{R}^{m \times 3}$.
 for $k = 0, 1, \ldots, p$
  Determine $x_{k+1} = \text{locmin}\,(\langle f \rangle_{\lambda_k}, x_k)$.
 end do

The vector x_{p+1} is a candidate for the global minimizer. In most cases we use gmin with a set of randomly generated starting points with $p > 0$.

Setting $p = 0$ in gmin reduces to the use of locmin on the original function f from a random starting point. A standard multistart method is obtained if gmin is used from a set of randomly generated starting points with $p = 0$.

Algorithm gmin depends on the optimization procedure $\text{locmin}(\cdot, \cdot)$ and on the choice of the continuation parameters λ_k. We will discuss the choice of optimization procedure later; in our computational experiments the continuation parameters are determined by setting

$$\lambda_k = \left(1 - \frac{k}{p}\right)\lambda_0.$$

More sophisticated choices that make use of the behavior of $\langle f \rangle_\lambda$ along the path are clearly possible.

The molecular optimization problems that we are considering can be modeled in terms of the potential function

$$(4.1) \qquad f(x) = \sum_{i,j \in \mathcal{S}} p_{i,j}(x_i - x_j),$$

where $p_{i,j} : \mathbb{R}^3 \mapsto \mathbb{R}$ is the pairwise potential. Algorithm gmin can be used to determine the global minimizer of f once we determine the Gaussian transform $\langle f \rangle_\lambda$. We have already noted that Theorem 3.3 implies that

$$(4.2) \qquad \langle f \rangle_\lambda(x) = \sum_{i,j \in \mathcal{S}} \langle p_{i,j} \rangle_{\sqrt{2}\lambda}(x_i - x_j).$$

Hence, we need to determine the Gaussian transform of $p_{i,j}$.

For the distance geometry problem (2.1), the pairwise potential $p_{i,j}$ is defined by (3.5). This potential is decomposable, and the Gaussian transform of $p_{i,j}$ is given by (3.6). Hence, (4.2) shows that the Gaussian transform for the distance geometry problem (2.1) is

$$(4.3) \quad \langle f \rangle_\lambda(x) = \sum_{(i,j) \in \mathcal{S}} \left[(\|x_i - x_j\|^2 - \delta_{i,j}^2)^2 + 10\lambda^2 \|x_i - x_j\|^2 \right] + \gamma,$$

where

$$\gamma = \sum_{(i,j) \in \mathcal{S}} \left(15\lambda^4 - 6\delta_{i,j}^2 \lambda^2 \right).$$

For the distance geometry problem (2.2), the pairwise potential is

$$p_{i,j}(x) = \min^2 \left\{ \frac{\|x\|^2 - l_{i,j}^2}{l_{i,j}^2}, 0 \right\} + \max^2 \left\{ \frac{\|x\|^2 - u_{i,j}^2}{u_{i,j}^2}, 0 \right\}.$$

In this case, the potential $p_{i,j}$ is not decomposable, but $p_{i,j}(x) = h_{i,j}(\|x\|)$, where

$$(4.4) \qquad h_{i,j}(r) = \min^2 \left\{ \frac{r^2 - l_{i,j}^2}{l_{i,j}^2}, 0 \right\} + \max^2 \left\{ \frac{r^2 - u_{i,j}^2}{u_{i,j}^2}, 0 \right\}.$$

Since $h_{i,j}$ is an even function, Theorem 3.4 shows that

$$
\begin{aligned}
\langle p_{i,j} \rangle_\lambda(x) &= \frac{1}{\lambda\sqrt{\pi}\,r} \int_{-\infty}^{+\infty} s h_{i,j}(s) \exp\left(-\frac{(r-s)^2}{\lambda^2}\right) ds \\
&= \frac{1}{\sqrt{\pi}\,r} \int_{-\infty}^{+\infty} (r+\lambda s) h_{i,j}(r+\lambda s) \exp\left(-s^2\right) ds,
\end{aligned}
$$

where $r = \|x\|$, and thus (4.2) yields that the Gaussian transform for the distance geometry problem (2.2) is

$$
\langle f \rangle_\lambda(x) = \sum_{i,j \in \mathcal{S}} \frac{1}{\sqrt{\pi}\,r_{i,j}} \int_{-\infty}^{+\infty} (r_{i,j} + \sqrt{2}\lambda s) h_{i,j}(r_{i,j} + \sqrt{2}\lambda s) \exp\left(-s^2\right) ds,
$$

where $r_{i,j} = \|x_i - x_j\|$. In our computational experiments we use the Gauss-Hermite approximation

$$
(4.5) \quad \langle f \rangle_{\lambda,q}(x) = \sum_{i,j \in \mathcal{S}} \frac{1}{\sqrt{\pi}\,r_{i,j}} \sum_{k=1}^{q} w_k(r_{i,j} + \sqrt{2}\lambda s_k) h_{i,j}(r_{i,j} + \sqrt{2}\lambda s_k),
$$

where w_k and s_k are the weights and nodes for the Gaussian quadrature, respectively

The functions defined by (4.3) and (4.5) are partially separable because $\langle f \rangle_\lambda$ and $\langle f \rangle_{\lambda,q}$ are the sum of $|\mathcal{S}|$ functions that depend on six variables. We note that the number of flops required to compute the function, gradient, or Hessian matrix of $\langle f \rangle_\lambda$ and $\langle f \rangle_{\lambda,q}$ is of order $|\mathcal{S}|$ because the function and derivatives of each element function can be evaluated with a constant number of flops. An important difference between (4.3) and (4.5), with $h_{i,j}$ defined by (4.4), is that $\langle f \rangle_\lambda$ is infinitely differentiable for any $\lambda \geq 0$, while $\langle f \rangle_{\lambda,q}$ is only continuously differentiable with a piecewise continuous Hessian matrix $\nabla^2 \langle f \rangle_{\lambda,q}$. This difference affects the choice of the procedure locmin.

In discussing the choice of locmin, we assume that we are dealing with distance geometry problems, or more generally, with a problem where $|\mathcal{S}|$ is bounded by a small multiple of n. In this case $\langle f \rangle_\lambda$ and $\langle f \rangle_{\lambda,q}$ have sparse Hessian matrices, so it makes sense to take advantage of this structure. Problems where $|\mathcal{S}|$ is essentially n^2 (for example, cluster problems) are discussed in the next section.

A Newton method that takes into account the sparsity of the problem is probably the best choice for locmin if the function is defined by (4.3) because for these problems the cost of function, gradient, and Hessian matrix evaluation is of order n, and the cost per iteration is also of order n. Moré and Wu [38] used a trust region version because these algorithms are able to escape regions of negative curvature that are present in these problems.

The choice of procedure locmin has to be done with some care for the function defined by (4.5) because $\langle f \rangle_{\lambda,q}$ is not twice continuously differentiable. The Hessian matrix is discontinuous at points where the argument of $h_{i,j}$ coincides with either $l_{i,j}$ or $u_{i,j}$. We cannot expect to avoid these discontinuities, in particular, if $l_{i,j}$ or $u_{i,j}$ are close. Moré and Wu [37] used the variable-metric limited-memory code vmlm in MINPACK-2, which is an implementation of the Liu and Nocedal [36] algorithm.

The formulation of the distance geometry problem (2.2) in terms of $p_{i,j}(x) = h_{i,j}(\|x\|)$ where $h_{i,j}$ is defined by (4.4) is typical, but other formulations have been used. Crippen and Havel [9] used

$$h_{i,j}(r) = \min^2 \left\{ \frac{r^2 - l_{i,j}^2}{r^2}, 0 \right\} + \max^2 \left\{ \frac{r^2 - u_{i,j}^2}{u_{i,j}^2}, 0 \right\},$$

because they felt that this formulation leads to a problem with fewer minimizers, but Havel [21] advocates the use of

$$h_{i,j}(r) = \min^2 \left\{ \frac{r^2 - l_{i,j}^2}{r^2 + l_{i,j}^2}, 0 \right\} + \max^2 \left\{ \frac{r^2 - u_{i,j}^2}{u_{i,j}^2}, 0 \right\}$$

because this formulation avoids the barrier created at $r = 0$. In both formulations $p_{i,j}$ has a discontinuous second derivative. If we use

$$(4.6) \qquad h_{i,j}(r) = \begin{cases} \left(\dfrac{l_{i,j}^2 - r^2}{l_{i,j}^2} \right)^3 & \text{if} \quad r < l_{i,j}, \\ 0 & \text{if} \quad r \in [l_{i,j}, u_{i,j}], \\ \left(\dfrac{r^2 - u_{i,j}^2}{u_{i,j}^2} \right)^3 & \text{if} \quad r > u_{i,j}, \end{cases}$$

then $p_{i,j}$ is twice continuously differentiable. Moreover, if f is defined by (4.1), then $f(x) \geq 0$ with $f(x) = 0$ if and only if x solves the distance geometry problem (2.2). From an optimization point of view, formulation (4.6) is preferable because it allows the use of a Newton method in locmin.

The computational experiments performed by Moré and Wu [38,37] on various distance geometry problems show that algorithm gmin is able to find global minimizers reliably and efficiently. An interesting aspect of these results is that algorithm gmin with $p > 0$ requires less than twice the effort (measured in terms of function and gradient evaluations) than $p = 0$. At first sight this is surprising because gmin with $p > 0$ requires the solution of p minimization problems. However, for reasonable choices of λ_0, finding a minimizer of $\langle f \rangle_\lambda$ with $\lambda = \lambda_0$ is found quickly because $\langle f \rangle_\lambda$ is a smooth, well-behaved function. A minimizer of $\langle f \rangle_\lambda$ with $\lambda = \lambda_k$ is also found quickly because x_k is a good starting point. On the other hand, gmin with $p = 0$ must find a local minimizer of f, which is not necessarily smooth or well-behaved, from a starting point that is not guaranteed to be near a minimizer. We expect that future work will improve the continuation procedure and further reduce the cost of the continuation procedure.

5. Future directions. Global smoothing and continuation have proved to be effective tools for the solution of molecular optimization problems with a moderate number of atoms, but improvements in these techniques will be needed to address problems with a large number of atoms. In this section we outline possible extensions to the work that we have presented.

5.1. Continuation algorithms. Algorithm gmin is a relatively simple algorithm for tracing a curve $x(\lambda)$, where $x(\lambda)$ is a minimizer of $\langle f \rangle_\lambda$. For problems with a large number of atoms we need to improve gmin by computing $x(\lambda)$ more efficiently. If we define function $h : \mathbb{R}^n \times \mathbb{R} \mapsto \mathbb{R}$ by

$$h(x, \lambda) = \langle f \rangle_\lambda(x),$$

and differentiate twice with respect to the variable x, we obtain

$$\partial_{xx} h[x(\lambda), \lambda] x'(\lambda) + \partial_{\lambda x} h[x(\lambda), \lambda] = 0.$$

This differential equation, together with the initial value $x(0) = x_0$, defines a trajectory $x(\lambda)$ under suitable nondegeneracy assumptions. We can use continuation algorithms (see, for example, Allgower and Georg [2]), but these algorithms are designed to trace stationary points of h, that is, solutions to

$$\partial_x h[x(\lambda), \lambda] = 0.$$

Our situation is somewhat different because we need to trace minimizers of h. However, in general it is not possible to define a continuous trajectory of minimizers, and thus we must be prepared to jump curves. For additional information on issues related to tracing minimizers, see Gudat, Guerra Vazquez, and Jongen [18].

5.2. Smoothing. The Gaussian transform is isotropic because if we view the function f in a different coordinate system via the function $h : \mathbb{R}^n \mapsto \mathbb{R}$ defined by

$$h(x) = f(P^T x),$$

then $\langle h \rangle_\lambda(Px) = \langle f \rangle_\lambda(x)$ for any orthogonal matrix $P \in \mathbb{R}^{n \times n}$. If we wish to emphasize some directions, we can generalize the Gaussian transform to the *anisotropic* transform,

$$\langle f \rangle_\Lambda(x) = \frac{1}{\pi^{n/2} |\det \Lambda|} \int_{\mathbb{R}^n} f(y) \exp\left(-\|\Lambda^{-1}(y - x)\|^2\right) \, dy,$$

where Λ is a nonsingular matrix. For this transformation $\langle h \rangle_\Lambda (Px) = \langle f \rangle_\Lambda (x)$ if $P\Lambda$ is orthogonal, so that the scaling in Λ controls the smoothing. Orešič and Shalloway [41] and Straub, Ma, and Arena [58] have used the anisotropic Gaussian transform for molecular optimization. Wu [60] showed that if Λ is a diagonal matrix, then the anisotropic transform can be applied in the same way as the isotropic transform to decomposable functions and potential functions in molecular optimization problems.

The Gaussian transform can also be extended to a general density function $\rho : \mathbb{R}^n \mapsto \mathbb{R}$ by defining the *generalized* transform by

$$\langle\!\langle f \rangle\!\rangle_\lambda(x) = \frac{1}{\lambda^n} \int_{\mathbf{R}^n} f(y)\, \rho\left(\frac{x-y}{\lambda}\right)\, dy.$$

The analysis of Moré and Wu [39] shows that the smoothing properties of the Gaussian transform can be extended to this class of transformation. By admitting a larger class of transformations, we should be able to extend the range of functions that can be transformed.

5.3. Optimization algorithms. Newton methods are appropriate for distance geometry problems where $|\mathcal{S}|$ is of order n because for these problems the cost of function, gradient, and Hessian matrix evaluation is of order n, and the cost per iteration is also of order n. A standard Newton method is not appropriate for large cluster problems where $|\mathcal{S}|$ is of order n^2 because the storage is of order n^2 and the cost per iteration is of order n^3.

The limited-memory variable-metric method [36] is suitable for systems with a large number of atoms because the memory requirements and cost per iteration is of order n. Unfortunately, the number of iterations required for convergence on these problems increases rapidly with the number of atoms. Preliminary experiments with a standard truncated Newton method showed that this method required a large number of inner conjugate gradient iterations. Since each conjugate gradient iteration requires order n operations, it is not surprising that the standard truncated Newton method required more computing time than the limited-memory variable-metric method.

We expect that a truncated Newton method with a suitable preconditioner will reduce the computing time required to solve cluster problems. Schlick and Fogelson [51,52] developed such an algorithm for molecular dynamics simulation and structure refinement, with a preconditioner constructed from an approximate Hessian matrix. Similar ideas should apply to cluster problems.

5.4. Function evaluations. We have already noted that the number of flops required to compute the function and derivatives in a distance geometry problem is of order $|\mathcal{S}|$. In distance geometry problems $|\mathcal{S}|$ is of order n, and thus we can evaluate these functions in order n. In cluster problems, however, all pairwise potentials are included, and then the cost of evaluating the function, gradient, and Hessian matrix is of order n^2. This represents a major hurdle to the solution of large cluster problems, since in a typical problem we need multiple runs and hundreds of function evaluations per run to determine the global minimizer.

We can reduce the cost of the function evaluation by computing an approximation to the function. The fast multipole method (Greengard and Rokhlin [16], and Greengard [15]), in particular, has attracted considerable attention because the running time is proportional to n. However, implementation of the fast multipole method requires considerable care and analysis, so only sophisticated implementations are able to achieve the order n running time.

Board and coworkers [35,34,45] have developed several sequential and parallel packages for computing electrostatic force fields and potentials using fast multipole algorithms. These implementations have been done with considerable care but are geared to molecular dynamics simulations where it is reasonable to assume a uniform distribution of atoms. The performance of these algorithms degrades considerably in an optimization setting because the distribution of the atoms is not uniform, unless we are in the final stages of convergence. In our opinion the only currently effective method for reducing the computing time of the function evaluation in large cluster problems is to evaluate the function in parallel.

5.5. High-performance architectures. Macromolecular systems usually have 1,000 to 10,000 atoms, and the ionic systems of physical interest, as we have mentioned, may contain as many as 200,000 ions. Determining the global solutions for these problems will not be feasible without the use of parallel high-performance architectures, even with the most efficient optimization algorithm.

The global continuation algorithm can be parallelized easily at a coarse level, with each processor assigned the computation of a solution trajectory. This strategy requires little communication among processors and is suitable for massively parallel architectures, such as the IBM SP. Indeed, we have implemented, for example, the continuation algorithms for distance geometry problems on the IBM SP at Argonne. Although we have not yet tested the algorithms with large problems, the results on medium-sized problems (with 500 atoms) show that the algorithms have satisfactory performance on as many as 64 processors.

Load balancing and synchronization between the processors are two of the problems that must be addressed for systems with a large number of atoms. Load balancing can be a problem because trajectories may require different amount of computing time. Another problem is that processors may end up tracing the same solution trajectory, even if they are given different starting points. Synchronization between the processors will be required to make sure that different trajectories are traced.

Acknowledgments. Our research on macromolecular global optimization problems has been influenced by John Schiffer's work on ionic systems and Julius Jellinek's work on clusters. Steve Pieper deserves special mention for bringing the work on ionic systems to our attention and for sharing his insights on this problem. We are also grateful to Ian Coope for pointing out that the distance geometry problem (2.1) can be solved as a linear algebra problem if all the pairwise distances are available. Gail Pieper's incisive comments on the manuscript are also gratefully acknowledged. We also thank the anonymous referee for helpful comments and suggestions that improved the manuscript.

REFERENCES

[1] S. AL-HOMIDAN AND R. FLETCHER, *Hybrid methods for finding the nearest Euclidean distance matrix*, technical report, The University of Dundee, Dundee, Scotland, 1995.

[2] E.L. ALLGOWER AND K. GEORG, *Numerical Continuation: An Introduction*, Springer-Verlag, 1990.

[3] L.M. BLUMENTHAL, *Theory and Applications of Distance Geometry*, Oxford University Press, 1953.

[4] A.T. BRÜNGER AND M. NILGES, *Computational challenges for macromolecular structure determination by X-ray crystallography and solution NMR-spectroscopy*, Q. Rev. Biophys., 26 (1993), pp. 49–125.

[5] R.H. BYRD, E. ESKOW, AND R.B. SCHNABEL, *A new large-scale global optimization method and its application to Lennard-Jones problems*, Technical report CU-CS-630-92, Department of Computer Science, University of Colorado, Boulder, Colorado, revised, 1995.

[6] R.H. BYRD, E. ESKOW, R.B. SCHNABEL, AND S.L. SMITH, *Parallel global optimization: Numerical methods, dynamic scheduling methods, and application to molecular configuration*, in Parallel Computation, B. Ford and A. Fincham, eds., Oxford University Press, 1993, pp. 187–207.

[7] T.F. COLEMAN, D. SHALLOWAY, AND Z. WU, *Isotropic effective energy simulated annealing searches for low energy molecular cluster states*, Comp. Optim. Applications, 2 (1993), pp. 145–170.

[8] ———, *A parallel build-up algorithm for global energy minimizations of molecular clusters using effective energy simulated annealing*, J. Global Optim., 4 (1994), pp. 171–185.

[9] G.M. CRIPPEN AND T.F. HAVEL, *Distance Geometry and Molecular Conformation*, John Wiley & Sons, 1988.

[10] Y.M. ERMOLIEV, V.I. NORKIN, AND R.J.-B. WETS, *The minimization of discontinuous functions: Mollifier subgradients*, SIAM J. Control Optim., 33 (1995), pp. 149–167.

[11] M.R. GAREY AND D.S. JOHNSON, *Computers and Intractability*, W. H. Freeman, 1979.

[12] W. GAUTSCHI, *Algorithm 726: ORTHOPOL – A package of routines for generating orthogonal polynomials and Gauss-type quadrature rules*, ACM Trans. Math. Software, 20 (1994), pp. 21–62.

[13] W. GLUNT, T.L. HAYDEN, AND M. RAYDAN, *Molecular conformation from distance matrices*, J. Comp. Chem., 14 (1993), pp. 114–120.

[14] ———, *Preconditioners for distance matrix algorithms*, J. Comp. Chem., 15 (1994), pp. 227–232.

[15] L. GREENGARD, *The Rapid Evaluation of Potential Fields in Particle Systems*, MIT Press, 1988.

[16] L. GREENGARD AND V. ROKHLIN, *A fast algorithm for particle simulation*, J. Comput. Phys., 73 (1987), pp. 325–348.

[17] A. GRIEWANK, *Generalized descent for global optimization*, J. Optim. Theory Appl., 34 (1981), pp. 11–39.

[18] J. GUDDAT, F.G. VAZQUEZ, AND H.T. JONGEN, *Parametric Optimization: Singularities, Pathfollowing and Jumps*, John Wiley & Sons, 1990.

[19] H. HABERLAND, ed., *Clusters of Atoms and Molecules, Springer Series in Chemical Physics*, vol. 52, Springer-Verlag, 1994.

[20] R.W. HASSE AND J.P. SCHIFFER, *The structure of the cylindrically confined coulomb lattice*, Ann. Physics, 203 (1990), pp. 419–448.

[21] T.F. HAVEL, *An evaluation of computational strategies for use in the determination of protein structure from distance geometry constraints obtained by nuclear magnetic resonance*, Prog. Biophys. Mol. Biol., 56 (1991), pp. 43–78.

[22] B.A. HENDRICKSON, *The molecule problem: Determining conformation from pairwise distances*, PhD thesis, Cornell University, Ithaca, New York, 1991.

[23] ———, *The molecule problem: Exploiting structure in global optimization*, SIAM J. Optimization, 5 (1995), pp. 835–857.

[24] M.R. HOARE, *Structure and dynamics of simple microclusters*, Advances in Chemical Physics, 40 (1979), pp. 49–135.

[25] M.R. HOARE AND J. McINNES, *Statistical mechanics and morphology of very small atomic clusters*, Faraday Discuss. Chem. Soc., 61 (1976), pp. 12–24.

[26] M.R. HOARE AND P. PAL, *Statistics and stability of small assemblies of atoms*, J. Cryst. Growth, 17 (1972), pp. 77–96.

[27] J. JELLINEK, *Theoretical dynamical studies of metal clusters and cluster-ligand systems*, in Metal-Ligand Interactions: Structure and Reactivity, N. Russo, ed., Kluwer Academic Publishers, 1995 (in press).

[28] V.Y. KATKOVNIK AND O.Y. KULCHITSKII, *Convergence of a class of random search algorithms*, Automat. Remote Control, 8 (1972), pp. 81–87.

[29] J. KOSTROWICKI AND L. PIELA, *Diffusion equation method of global minimization: Performance for standard functions*, J. Optim. Theory Appl., 69 (1991), pp. 269–284.

[30] J. KOSTROWICKI, L. PIELA, B.J. CHERAYIL, AND H.A. SCHERAGA, *Performance of the diffusion equation method in searches for optimum structures of clusters of Lennard-Jones atoms*, J. Phys. Chem., 95 (1991), pp. 4113–4119.

[31] J. KOSTROWICKI AND H.A. SCHERAGA, *Application of the diffusion equation method for global optimization to oligopeptides*, J. Phys. Chem., 96 (1992), pp. 7442–7449.

[32] J. KREIMER AND R.Y. RUBINSTEIN, *Nondifferentiable optimization via smooth approximation: General analytical approach*, Math. Oper. Res., 39 (1992), pp. 97–119.

[33] I.D. KUNTZ, J.F. THOMASON, AND C.M. OSHIRO, *Distance geometry*, in Methods in Enzymology, N. J. Oppenheimer and T. L. James, eds., vol. 177, Academic Press, 1993, pp. 159–204.

[34] C.G. LAMBERT AND J.A. BOARD, *A multipole-based algorithm for efficient calculation of forces and potentials in macroscopic periodid assemblies of particles*, Technical report 95-001, Department of Electrical Engineering, Duke University, Durham, North Carolina, 1995.

[35] J.F. LEATHRUM AND J.A. BOARD, *The parallel fast multipole algorithm in three dimensions*, Technical report TR92-001, Department of Electrical Engineering, Duke University, Durham, North Carolina, 1992.

[36] D.C. LIU AND J. NOCEDAL, *On the limited memory BFGS method for large scale optimization*, Math. Programming, 45 (1989), pp. 503–528.

[37] J.J. MORÉ AND Z. WU, *$\mathcal{E}$-optimal solutions to distance geometry problems via global continuation*, in Global Minimization of Nonconvex Energy Functions: Molecular Conformation and Protein Folding, P. M. Pardalos, D. Shalloway, and G. Xue, eds., American Mathemtical Society, 1995, pp. 151–168.

[38] ——, *Global continuation for distance geometry problems*, Preprint MCS-P505-0395, Argonne National Laboratory, Argonne, Illinois, 1995.

[39] ——, *Smoothing techniques for macromolecular global optimization*, Preprint MCS-P542-0995, Argonne National Laboratory, Argonne, Illinois, 1995.

[40] J.A. NORTHBY, *Structure and binding of Lennard-Jones clusters:* $13 \leq n \leq 147$, Journal of Chemical Physics, 87 (1987), pp. 6166–6177.

[41] M. OREŠIČ AND D. SHALLOWAY, *Hierarchical characterization of energy landscapes using Gaussian packet states*, J. Chem. Phys., 101 (1994), pp. 9844–9857.

[42] P.M. PARDALOS, D. SHALLOWAY, AND G. XUE, *Optimization methods for computing global minima of nonconvex potential energy functions*, J. Global Optim., 4 (1994), pp. 117–133.

[43] L. PIELA, J. KOSTROWICKI, AND H.A. SCHERAGA, *The multiple-minima problem in the conformational analysis of molecules: Deformation of the protein energy hypersurface by the diffusion equation method*, J. Phys. Chem., 93 (1989), pp. 3339–3346.

[44] R. RAFAC, J.P. SCHIFFER, J.S. HANGST, D.H.E. DUBIN, AND D.J. WALES, *Stable configurations of confined cold ionic systems*, Proc. Natl. Acad. Sci. U.S.A., 88 (1991), pp. 483–486.

[45] W.T. RANKIN AND J.A. BOARD, *A portable distributed implementation of the parallel multipole tree algorithm*, Technical report 95-002, Department of Electrical Engineering, Duke University, Durham, North Carolina, 1995.

[46] P.J. REYNOLDS, ed., *On Clusters and Clustering*, North-Holland, 1993.

[47] R.Y. RUBINSTEIN, *Smoothed functionals in stochastic optimization*, Math. Oper. Res., 8 (1983), pp. 26–33.

[48] J.B. SAXE, *Embeddability of weighted graphs in k-space is strongly NP-hard*, in Proc. 17th Allerton Conference in Communications, Control and Computing, 1979, pp. 480–489.

[49] H.A. SCHERAGA, *Predicting three-dimensional structures of oligopeptides*, in Reviews in Computational Chemistry, K. B. Lipkowitz and D. B. Boyd, eds., vol. 3, VCH Publishers, 1992, pp. 73–142.

[50] J.P. SCHIFFER, *Phase transitions in anisotropically confined ionic crystals*, Phys. Rev. Lett., 70 (1993), pp. 818–821.

[51] T. SCHLICK AND A. FOGELSON, *TNPACK – A truncated Newton minimization package for large-scale problems: I. Algorithms and usage*, ACM Trans. Math. Software, 18 (1992), pp. 46–70.

[52] ——, *TNPACK – A truncated Newton minimization package for large-scale problems: II. Implementations examples*, ACM Trans. Math. Software, 18 (1992), pp. 71–111.

[53] D. SHALLOWAY, *Application of the renormalization group to deterministic global minimization of molecular conformation energy functions*, J. Global Optim., 2 (1992), pp. 281–311.

[54] ——, *Packet annealing: A deterministic method for global minimization, application to molecular conformation*, in Recent Advances in Global Optimization, C. Floudas and P. Pardalos, eds., Princeton University Press, 1992, pp. 433–477.

[55] N,M. STEEN, G.D. BYRNE, AND E.M. GELBARD, *Gaussian quadratures for the integrals* $\int_0^\infty \exp(-x^2) f(x)\, dx$ *and* $\int_0^b \exp(-x^2) f(x)\, dx$, Math. Comp., 23 (1969), pp. 661–674.

[56] F.H. STILLINGER, *Role of potential-energy scaling in the low-temperature relaxation behavior of amorphous materials*, Phys. Rev. B, 32 (1985), pp. 3134–3141.

[57] J.E. STRAUB, *Optimization techniques with applications to proteins*, preprint, Boston University, Department of Chemistry, Boston, Massachusetts, 1994.

[58] J.E. STRAUB, J. MA, AND P. AMARA, *Simulated anealing using coarse-grained clasical dynamics: Fokker-Planck and Smoluchowski dynamics in the Gaussian density approximation*, J. Chem. Phys., 103 (1995), pp. 1574–1581.

[59] A.H. STROUD AND D. SECREST, *Gaussian Quadrature Formulas*, Prentice-Hall, Inc., 1966.

[60] Z. WU, *The effective energy transformation scheme as a special continuation approach to global optimization with application to molecular conformation*, Preprint MCS-P442-0694, Argonne National Laboratory, Argonne, Illinois, 1994.

[61] G.L. XUE, *Improvement on the Northby algorithm for molecular conformation: Better solutions*, J. Global. Optim., 4 (1994), pp. 425–440.

GLOBAL MINIMIZATION OF LENNARD-JONES FUNCTIONS ON TRANSPUTER NETWORKS

KLAUS RITTER*, STEPHEN M. ROBINSON†, AND STEFAN SCHÄFFLER*

Abstract. This paper presents a three-phase computational procedure for minimizing molecular energy potential functions of the pure Lennard-Jones type. The first phase consists of a special heuristic for generating an initial atomic configuration. In the second phase a global minimization method is applied to compute a configuration close to the optimal solution. Finally, the third phase uses this configuration as the starting point for a local minimization method. Since the second and third phases are very suitable for parallel implementation, we describe briefly our implementation of the method in a parallel version of C on a transputer network, and we exhibit numerical results for approximate optimization of clusters with up to 20,000 atoms.

Key words. global optimization, Lennard-Jones potential, transputer networks, parallel computing.

AMS(MOS) 1980 subject classifications. 60G17, 60H10, 65K05, 90C27, 90C30.

1. Introduction. In this paper we consider the global minimization of the pure Lennard-Jones potential function $f : \mathbf{R}^{3N} \to \mathbf{R}$ for N atoms. Let $\xi_i \in \mathbf{R}^3$ denote the cartesian position of the ith atom in the cluster and let $n_{i,j}(x)$ be the squared distance between the ith and the jth atoms, *i.e.*,

$$n_{i,j}(x) := \| \xi_i - \xi_j \|^2 \quad \text{with} \quad x := \begin{pmatrix} \xi_1 \\ \vdots \\ \xi_N \end{pmatrix} \in \mathbf{R}^{3N};$$

then the Lennard-Jones potential function is given by

$$f(x) = \sum_{i=1}^{N-1} \sum_{j=i+1}^{N} (n_{i,j}^{-6}(x) - 2n_{i,j}^{-3}(x)).$$

The number of local minimizers of f has been conjectured to increase exponentially with N^2 [4]; for this reason the application of global minimization procedures that do not use the special structure of the objective function

---

* Fakultät für Mathematik, Technische Universität München, Arcisstr. 21, 80290 München, Germany.

† Department of Industrial Engineering, University of Wisconsin-Madison, 1513 University Avenue, Madison, WI 53706-1572, USA.

The research reported here was sponsored in part by the Air Force Systems Command, USAF, under Grant F49620-95-1-0222. The U. S. Government has certain rights in this material, and is authorized to reproduce and distribute reprints for Governmental purposes notwithstanding any copyright notation thereon. The views and conclusions contained herein are those of the authors and should not be interpreted as necessarily representing the official policies or endorsements, either expressed or implied, of the sponsoring agencies or the U. S. Government.

(*cf.* [1], [2], and [3]) can be very inefficient. For $N \leq 147$, Northby [7] considered some heuristic initial configurations based on the solution of very expensive discrete optimization problems. In some papers (see, *e.g.* [6]) Northby's approach is adopted for computations on massively parallel architectures. Nevertheless, the number N of atoms is restricted by the complexity of the discrete optimization problems.

In the remainder of this paper we propose a procedure for the global minimization of f, consisting of three phases. The first of these consists of a fast and apparently efficient heuristic for the computation of an initial configuration. This heuristic is described in Section 2 and is not based on the solution of a discrete optimization problem. Section 3 deals with the second phase of our procedure, in which we apply the global optimization method of [9] to the objective function f with the computed initial configuration as starting cluster in order to compute a configuration close to the optimal solution. In the final phase we use this new configuration as the starting point for a local minimization method; this procedure is developed in Section 4.

Since the second and third phases of the suggested procedure are well suited to implementation in parallel, we describe in Section 5 an implementation on a transputer network, using a parallel version of C. We present numerical results from this implementation showing approximate optimization of clusters with up to 20,000 atoms.

2. The Heuristic for the initial configuration. In this section we describe a heuristic procedure used to find an initial configuration to which we later apply a global minimization method. The special structure of the objective function f and its analysis in [7] lead to the application of icosahedral shells as principal schemes for the configurations of the atoms. An atom is placed on the ith shell if each shell up to and including the $(i-1)$st contains the maximal number of atoms. All shells have the same midpoint, which is referred to as the 0th shell and consists of the first atom. Each shell is of icosahedral form and is composed of 20 equilateral triangles.

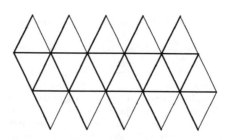

FIG. 1. *Lattice of an icosahedral shell*

Each triangle consists of a fixed number of atoms depending on the number of the shell. The distance between two neighboring atoms in a triangle is equal to 1. This choice is motivated by the special structure of f, which can be rewritten as

$$f(x) = \sum_{i=1}^{N-1} \sum_{j=i+1}^{N} [(n_{i,j}^{-3}(x) - 1)^2 - 1],$$

so that if it were possible one would want to have the distance between each pair of atoms in the entire cluster be equal to 1.

FIG. 2. *Atoms in a triangle on the first shell*

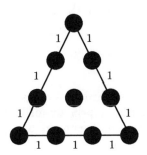

FIG. 3. *Atoms in a triangle on the third shell*

The maximal number of atoms on the first m shells is then given by

$$(2.1) \qquad 1 + 2m + \frac{5}{3}m(m+1)(2m+1).$$

The formula (2.1) takes into account the fact that one atom can be a member of different triangles.

Beginning with the first shell, we place atoms on a shell triangle by triangle, starting at the "north polar region" consisting of five triangles. Figure 5 shows the sequence in which the triangles are filled with atoms. Each atom in a triangle is placed in such a way that the distance between the new atom and the existing atoms in the triangle is equal to one as often as possible.

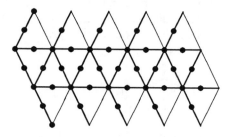

FIG. 4. *All atoms on the second shell*

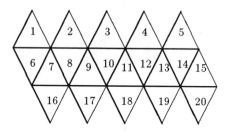

FIG. 5. *Sequence for filling triangles*

The approach we have described here yields a fast, practical method for computing an initial cluster x_0, especially for large N. This initial cluster is then used as the starting point for the global optimization method that we describe in the next section.

3. Global minimization. In this section we describe a global minimization method proposed in [9]. We shall use this method to improve the initial configuration obtained by the shelling procedure described in the last section.

Let $g : \mathbf{R}^n \to \mathbf{R}$ be a twice continuously differentiable objective function to be minimized. The procedure of [9] employs the following Itô stochastic differential equation; for detailed information about such Itô equations see [5] and [8].

$$(3.1) \qquad dX_t = -\nabla g(X_t)dt + \epsilon dB_t, \qquad X_0 = x_0,$$

where x_0 is a chosen starting point and $\{B_t\}_{t\geq 0}$ denotes n-dimensional Brownian motion. We summarize below some important properties of the above Itô differential equation, which hold for all C^2 functions g satisfying the following assumption, which determines the growth behavior of g for $\|x\| \to \infty$. In particular, it implies the standard "outwardness" condition often used to establish the existence of a solution (in this case, to $\nabla g(x) = 0$).

ASSUMPTION 3.1. *There exist positive real numbers ϵ and r such that for each $x \in \mathbf{R}^n$ with $\|x\| > r$,*

$$x^T \nabla g(x) \geq \frac{1}{2}(1 + n\epsilon^2) \max\{1, \|\nabla g(x)\|\}.$$

Evidently any function satisfying Assumption 3.1 has a global minimizer, but the converse is not true. Under this assumption one can establish the following result, in which the phrase "almost every" is used in the sense of n-dimensional Wiener measure. For further discussion and proofs we refer to [9].

THEOREM 3.1. *Let g be a real-valued C^2 function on $\mathbf{R}^n$ which, together with some positive ϵ and r, satisfies Assumption 3.1. Let x_0 be any point of $\mathbf{R}^n$. Then:*

a. There exists an unique stochastic process $\{X_t\}_{t \geq 0}$ solving (3.1).

b. Almost every path $\mathcal{P}$ of $\{X_t\}_{t \geq 0}$ has the following property: for any positive η, there is a finite time $T_{\mathcal{P}}$ within which the path $\mathcal{P}$ passes within a distance η of a global minimizer of g.

c. The random variables X_t, $t \geq 0$, converge in distribution to a random variable X whose distribution is given by the probability density function

$$p(x) = K \exp(-2\epsilon^{-2}g(x)),$$

where K is chosen so that $\int_{\mathbf{R}^n} p(x)dx = 1$. This density function p is independent of x_0 and takes its global maximum at exactly those points where the objective function g takes its global minimum.

For the numerical computation of a path of $\{X_t\}_{t \geq 0}$ we consider the following iteration scheme, which results from a standard approach in the numerical analysis of ordinary differential equations (semi-implicit Euler method with incomplete Jacobi matrix). For a fixed stepsize σ, we let $H(x)$ be the Hessian of g at x and set

$$(3.2) \quad x_{j+1}^1 := x_j - \Big[I_n + \sigma \mathrm{diag}\, H(x_j)\Big]^{-1}\Big[\sigma \nabla g(x_j) - \epsilon n_3(\sigma/2)^{1/2}\Big],$$

$$(3.3) \quad x(\sigma/2) := x_j - \Big[I_n + (\sigma/2)\mathrm{diag}\, H(x_j)\Big]^{-1}\Big[(\sigma/2)\nabla g(x_j) - \epsilon n_1(\sigma/2)^{1/2}\Big],$$

$$(3.4) \quad \begin{aligned} x_{j+1}^2 := x(\sigma/2) - &\Big[I_n + (\sigma/2)\mathrm{diag}\, H\big(x(\sigma/2)\big)\Big]^{-1} \\ &\times \Big[(\sigma/2)\nabla g\big(x(\sigma/2)\big) - \epsilon n_2(\sigma/2)^{1/2}\Big], \end{aligned}$$

where n_1 and n_2 are realizations of independent $N(0, I_n)$ normally distributed random vectors, which are computed by pseudo-random numbers, and $n_3 = n_1 + n_2$.

For a fixed positive δ we take $x_{j+1} = x_{j+1}^2$ if $\|x_{j+1}^1 - x_{j+1}^2\| \leq \delta$; otherwise steps (3.2) and (3.4) have to be repeated with $\sigma/2$ instead of σ. If one of the matrices $I_n + \sigma \mathrm{diag}\, H(x_j)$ or $I_n + (\sigma/2)\mathrm{diag}\, H(x(\sigma/2))$ is not positive definite, $\sigma/2$ has to be used instead of σ. After the choice of a maximal number J of iterations we accept the point $x_j \in \{x_0, \ldots, x_J\}$ with the smallest function value $g(x_j)$ as an approximate optimal solution.

The use of the diagonal matrix $\mathrm{diag}\, H(x_j)$ instead of $H(x_j)$ permits application of this method to large-scale global optimization problems. The fact that the Lennard-Jones potential function f is not twice continuously differentiable is not crucial because f has poles of even order and the set of poles is such that the main properties of the method can be preserved. It is easy to see that the application of penalty functions depending on the number N of atoms leads to an objective function f_1 having the same global minimizers as f but satisfying Assumption 3.1 for some $r > 0$ and each $\epsilon > 0$: for example,

$$f_1 = f + (\|x\|^2 - N)_+^4.$$

The derivatives of f_1 of first and second order can be computed analytically. These derivatives were used for the numerical computations illustrated in this paper.

4. Local minimization. After the computation of a cluster close to the optimal solution we consider the local minimization of the Lennard-Jones potential function f. The simplest optimization procedure is the gradient method, for which the iteration scheme is defined by

$$x_{j+1} := x_j - \sigma_j \nabla f(x_j),$$

where x_j denotes the current iterate, x_0 the approximately optimal cluster computed by the global minimization procedure, and σ_j the stepsize. This stepsize is determined by a partial one-dimensional minimization using, for example, the Armijo-Goldstein rule. Because of the disadvantage of slow convergence, other methods are often often used in practice: for instance, Maier *et al.* [6] use a special version of the BFGS method. The best methods to handle problems of this kind are Newton-type procedures using the first and second derivatives of f. However, the large dimensionality of Lennard-Jones problems causes excessive work in the numerical computations as well as the use of an enormous amount of storage.

On the other hand, our problems appear to concentrate the relevant information from the Hessian matrix on the diagonal elements. This leads us to employ the following iteration scheme, in which D_j denotes the diagonal of the Hessian matrix of f at x_j. If D_j is positive definite, then we take

$$x_{j+1} = x_j - \sigma_j D_j^{-1} \nabla f(x_j);$$

otherwise, we use the gradient procedure

$$x_{j+1} = x_j - \sigma_j \nabla f(x_j)$$

(that is, we replace D_j by the diagonal of the identity). The stepsize σ_j is chosen as mentioned above. The stopping criterion that we use takes x_{j+1} to be the optimal solution if

$$(f(x_{j+1}) - f(x_j))/(f(x_j) \cdot \sigma_j) < 10^{-7},$$

which again is motivated by the special structure of the problem. With this procedure we have been able to obtain fast convergence with small effort.

5. Implementation on transputer networks. In this section we discuss the application of transputer networks for the implementation of the mentioned procedure in a parallel version of C.

The transputer T805-25 from INMOS is a 32-bit microprocessor with a 64-bit floating point unit (FPU), graphics support, and 4 Kbytes on-chip RAM. This processor is able to perform 1.875 Mflops at a processor speed of 25MHz. A transputer has a configurable memory interface and four bidirectional communication links, which allow networks of transputers constructed by direct point-to-point connections without external logic. Each link can transfer data at rates up to 2.35 Mbytes/sec. The processor has access to a maximum of 4Gbytes of memory via the external memory interface. In a transputer the FPU, the central processing unit (CPU), and the data transfer work physically in parallel.

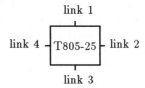

Fig. 6. *Transputer T805-25*

For the transputer architecture just described, it is advantageous to interpret the software as a set of processes that can be performed in serial or in parallel. After starting execution, a process performs a number of actions and then either terminates or pauses. Each action may be an assignment, an input, or an output. The communication between two processes is achieved by means of channels, where a channel between processes executing on the same transputer is implemented by a single word in the memory and a channel between different transputers is implemented by point-to-point links.

Transputers can be programmed in most high level languages. Using a parallel version of C it is possible to exploit the advantages of the transputer architecture. This version can be used to program an individual transputer or a network of transputers. When parallel C is used to program an individual transputer, the transputer shares its time between the concurrent processes and channel communication is implemented by moving data within the memory. Communication between processes on different transputers is implemented directly by transputer links. Since each transputer has four bidirectional links, one can conveniently use trinary trees as topologies for transputer networks.

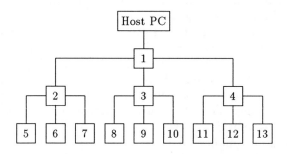

FIG. 7. *Trinary tree of depth 3 with 13 transputers*

Using the network of figure 7 we outline how one can parallelize the procedure given above. First of all, the number $n = 3N$ of variables has to be divided into 13 parts (corresponding to the 13 transputers), where each part gets nearly the same number of variables. Thus, each transputer is assigned a particular set of variables. Furthermore, each transputer has the formulas for the corresponding components of the gradient and the diagonal elements of the Hessian matrix in its local memory. After sending the current iterate x_j to all transputers, each transputer computes the corresponding components of the gradient and the Hessian matrix and sends the data back to the first transputer. These computations are done in parallel. The first transputer checks the stopping condition, computes the new iterate, and sends the results to the host PC.

6. Numerical results. In this section we exhibit some sample numerical results obtained by the procedures discussed above. For all of these computations we used an incomplete trinary tree network of depth

5 consisting of 95 transputers. A PC with an 80486 processor was used as host. The data given in the fourth column of the tables are the times in seconds needed for the local minimization (third phase of the method), which was the most time-consuming part of the algorithm. For the global optimization in the second phase we used a maximal iteration number $J = 50$, $\epsilon = 2$, $\delta = 0.1$, and the starting value for σ was equal to one.

We first show in Table 1 a set of results for atomic configurations with complete shells. Specifically, each row of Table 1 represents a configuration in which the number N of atoms is given by the formula (2.1) with the value of m shown in the first column of the table.

TABLE 1

Numerical results for complete shells up to the 18th shell

m	N	Transputers	L-J potential	Time
3	147	95	-8.76461e+02	23
4	309	95	-2.00722e+03	51
5	561	95	-3.84239e+03	112
6	923	95	-6.55272e+03	258
7	1415	95	-1.03089e+04	564
8	2057	95	-1.52815e+04	1082
9	2869	95	-2.16413e+04	2245
10	3871	95	-2.95588e+04	4229
11	5083	95	-3.92047e+04	7765
12	6525	95	-5.07496e+04	13420
13	8217	95	-6.43641e+04	23707
14	10179	95	-8.02189e+04	40005
15	12431	95	-9.84843e+04	63568
16	14993	95	-1.19331e+05	103865
17	17885	95	-1.42930e+05	160207
18	21127	95	-1.69452e+05	244795

In Table 2 we provide sample numerical results for atomic configurations with the number N of atoms ranging from $1,000$ to $20,000$.

The computed Lennard-Jones potentials shown here depend, of course, on the stopping criterion used in the algorithm. One might be able to find better potentials with different stopping criteria, and this is an interesting area for further study. However, we note that our results are very close to the best results published so far. Although the numerical examples

TABLE 2
Numerical results for N up to 20000

Size N	Transputers	L-J potential	Time
1000	95	-7.11780e+03	939
2000	95	-1.48269e+04	2135
3000	95	-2.26451e+04	4876
4000	95	-3.05479e+04	8000
5000	95	-3.85245e+04	11100
6000	95	-4.64918e+04	22109
7000	95	-5.45212e+04	26866
8000	95	-6.25840e+04	39409
9000	95	-7.06315e+04	48733
10000	95	-7.87336e+04	63324
20000	95	-1.60152e+05	304190

summarized in Table 1 and Table 2 are computed on a simple parallel architecture with only 95 processors, the results are comparable to those computed on large and powerful parallel architectures. This is possible since the three parts of the optimization procedure proposed in this paper are adapted especially to the optimization of Lennard-Jones functions.

REFERENCES

[1] R.H. BYRD, E. ESKOW, AND R.B. SCHNABEL, *A new large-scale global optimization method and its application to Lennard-Jones problems,* Technical report, University of Colorado at Boulder (1992).

[2] T. COLEMAN, D. SHALLOWAY, AND Z. WU, *Isotropic effective energy simulated annealing searches for low energy molecular cluster states,* Technical Report, Cornell Theory Center, Cornell University, Ithaca, NY (1992).

[3] T. COLEMAN, D. SHALLOWAY, AND Z. WU, *A parallel build-up algorithm for global energy minimization of molecular clusters using effective energy simulated annealing,*Technical Report, Cornell Theory Center, Cornell University, Ithaca, NY (1993).

[4] M.R. HOARE AND P. PAL, *Physical cluster mechanics: statics and energy surfaces for monatomic systems,*Adv. Phys. **20**, pp. 161–196 (1971).

[5] I. KARATZAS AND S.E. SHREVE, *Brownian Motion and Stochastic Calculus,* Springer-Verlag, New York, Berlin, Heidelberg (1991).

[6] R.S. MAIER, J.B. ROSEN, AND G.L. XUE, *A discrete-continuous algorithm for molecular energy minimization,*Technical report, U. S. Army High-Performance Computing Research Center, University of Minnesota, Minneapolis, MN (1992).

[7] J.A. NORTHBY, *Structure and binding of Lennard-Jones clusters: $13 \le N \le 147$,* J. Chem. Phys. **87**, pp. 6166–6177 (1987).

[8] P. PROTTER, *Stochastic Integration and Differential Equations*, Springer-Verlag, New York, Berlin, Heidelberg (1990).

[9] S. SCHÄFFLER, *Unconstrained global optimization using stochastic integral equations,Optimization* **35**, pp. 43–60 (1995).

VARIABLE-SCALE COARSE-GRAINING IN MACROMOLECULAR GLOBAL OPTIMIZATION

DAVID SHALLOWAY*

Abstract. We discuss the use of variable-scale coarse-graining for global minimization of macromolecular energy functions and compare three related methods. The Diffusion Equation Method and Adiabatic Gaussian Density Annealing are briefly reviewed and our Packet Annealing Method is discussed in more detail. All three methods dissect the global minimization problem into a sequence of local minimization problems on smoothed objective functions. They differ in the use of either energy averaging or free-energy averaging, the degree of anisotropy allowed in the objective function smoothing, and in tracking either single or multiple trajectories during the annealing procedure. Energy landscape scaling properties, which characterize the suitability of a potential landscape for this type of approach, are are also discussed.

Key words. Global minimization, protein folding, coarse-graining.

1. Introduction. Global minimization problems arise in many forms in studies of protein and macromolecular conformation [1]. Examples include structure-based drug design (predicting changes in protein and ligand structure upon binding) [2,3], experimental structural data refinement (from X-ray crystallography and multi-dimensional nuclear magnetic resonance) [4], refinement of structural predictions following homology-modeling [5,6], and prediction of the conformational changes induced by mutations or post-translational modifications. Predictive methods generally involve the minimization of objective functions that describe the potential energy $U(R)$ of a protein (often plus ligand) as a function over the space of all possible conformations parameterized by the multicomponent vector R which describes all the degrees-of-freedom in the system.

These problems are difficult because they have very large numbers of degrees of freedom ($\geq 10^4$ for many important problems), because the energy functions are extremely rough, and because they possess structure at widely-differing levels of spatial and energy scale. Many approaches have been tried [7], but none have been successful with the most important practical problems.

In our view, a flaw of many methods is that they have no systematic mechanism for coarse-graining to match the computational analysis to the dominant spatial scales of the problem at hand. For example, simulated annealing using molecular dynamics to sample conformation space is a popular global minimization method (e.g., see [4]), but its radius of convergence is severely limited.[1] This is because computational effort is

---

* Biophysics Program, Section of Biochemistry, Molecular and Cell Biology, Cornell University, Ithaca, NY 14853.

[1] For example, this approach can not practically predict the conformational changes that occur during ligand-protein binding. Even though it has been successful in X-ray crystallographic data refinement, an increased radius of convergence could enable

wasted in sampling the energy surface in excessive detail, i.e., at infinitesimal spatial scale. The same considerations pertain to standard Monte Carlo
sampling in which the energy function is evaluated at mathematical points
of infinitesimal size rather than over regions of finite size corresponding to
physical sampling.

In the past few years a number of groups have been developing methods that aim at accelerating the global minimization process by dissecting
it into a series of subproblems, each focusing on a different spatial scale.
Here we discuss some central issues pertaining to variable-scale coarse-
graining and compare three related methods. The *Diffusion Equation
Method* (DEM) developed by Piela, Kostrowicki, Scheraga and coworkers
and *Adiabatic Gaussian Density Annealing* (AGDA), developed by Straub
and coworkers are briefly reviewed; the *Packet Annealing Method* (PAM)
which is being developed by our group is discussed in more detail.

**1.1. Scale and the thermodynamic analogy in global mini-
mization.** A global minimum of objective function f over space S is typically described as a point

$$(1.1) \qquad x^* \in S \ni f(x^*) \leq f(x) \, \forall x \in S .$$

But this specification has no direct physical significance because it lacks
any measure of the relevant spatial or energy scales, i.e., how big a spatial
or energetic difference must be to be significant. Fig. 1 provides examples
of potential energy functions having minima that satisfy (1.1) but which
are not physically significant because either the depths or widths of the
minima are too small.

The narrow minimum marked with an arrow in panel (a) will not be of
physical significance if the statistical probability that the modeled physical
system will occupy the associated catchment region is negligible. On-the-
other-hand, the energy differences and barrier between the two minima in
panel (b) will be physically unimportant if these energies are much smaller
than the relevant energy scale. In that case the physically relevant global
minimum will lie between the two arrows. These considerations can be
systematically addressed by explicitly introducing appropriate scale parameters into the problem. The language of thermodynamics and statistical physics provides a natural vehicle for this task and a context for the
subsequent discussion.

We begin by reviewing some fundamental concepts [8]. For a molecule
in a thermal bath at temperature T, the Gibbs/Boltzmann probability
distribution $p_B(R)$ for occupation of an individual conformation R having
potential energy $U(R)$ is

$$(1.2) \qquad p_B(R) = \frac{e^{-\beta U(R)}}{\int e^{-\beta U(R)} \, dR} \equiv \frac{e^{-\beta U(R)}}{Z} .$$

---

lower-quality data to be used and reduce the frequency of misassignment errors.

The relevant energy scale is the thermal energy $\beta^{-1} \equiv k_B T$, where k_B is Boltzmann's constant. Proteins do not remain in single conformations but fluctuate through connected regions of conformation space. Each region corresponds to a *metastable macrostate*. Heuristically, we can think of each macrostate as a basin in the energy landscape which has been filled to a depth $\sim k_B T$. Thus, the sizes of the macrostates (which determine the relevant spatial scales of the problem) increase with temperature. The sizes depend on the potential landscape in a temperature-dependent manner and will vary between macrostates. The probability p_α that macrostate α is occupied is

$$(1.3) \qquad p_\alpha = \frac{\int_\alpha e^{-\beta U(R)} dR}{Z} \equiv \frac{e^{-\beta F_\alpha}}{Z} ,$$

where $\int_\alpha$ denotes integration over the macrostate region. This defines the *free-energy* F_α, a convenient measure of the occupation probability. F_α depends both on the *mean-energy* $\langle U \rangle_\alpha$ and *entropy* S_α of the macrostate. $\langle U \rangle_\alpha$ is calculated from

$$(1.4) \qquad \langle U \rangle_\alpha = \frac{\int_\alpha U(R) e^{-\beta U(R)} dR}{\int_\alpha e^{-\beta U(R)} dR} = \frac{\int_\alpha U(R) e^{-\beta U(R)} dR}{e^{-\beta F_\alpha}}$$

Roughly speaking, S_α is proportional to the logarithm of the volume in conformation space of the macrostate. A fundamental result is

$$(1.5) \qquad F_\alpha = \langle U \rangle_\alpha - T S_\alpha .$$

If the individual basins are approximated as quadratic, the integrals in (1.3) and (1.4) can be approximated as convolutions with Gaussians. We will see echoes of this in the DEM, AGDA and PAM.

These tools can be used to resolve the problems posed in Fig. 1. Rather than just measuring the depths of potential basins, we can calculate the probability contained within the basins at temperature T. This will depend both on their depths (i.e., U_α) and sizes (i.e., S_α) through Eqs. (1.3) and (1.5). Thus, from the physical standpoint, it is global minimization of the free-energy, not global minimization of the energy function itself, that is most relevant.

1.2. Variable scale potential smoothing and searching. The DEM, AGDA and PAM all seek global minima by performing multiple local minimizations on a sequence of transformed functions that have been smoothed at progressively smaller spatial scales. Although the motivations and mathematical derivations differ, in each method a smoothed objective function $f(R, \Lambda)$ is obtained by convolving the original objective function $f(R)$ with a Gaussian:

$$(1.6) \qquad f(R, \Lambda) = C \int f(R') e^{-\frac{1}{2}(R-R') \cdot \Lambda^{-2} \cdot (R-R')} dR' ,$$

where

$$(1.7) \qquad f(R) = \begin{cases} U(R) & \text{DEM and AGDA} \\ e^{-\frac{\beta}{2}U(R)} & \text{PAM} \end{cases}$$

and R represents all the conformation space variables. This smooths out the rapidly varying components of $f(R)$. The general properties of this class of transformation are discussed in Refs. [9] and [10]. In the DEM and AGDA the energy function itself is smoothed, while in the PAM the square-root of the Gibbs/Boltzmann distribution is smoothed. The spatial scale of smoothing is determined by Λ which, depending on the method, may be a scalar, a diagonal matrix, or a symmetric matrix. The PAM also includes the inverse temperature β as a parameter which determines the relevant energy scale.

The simplest search strategy consists of progressive local minimizations on a sequence of potentials smoothed to different spatial scales as illustrated in Fig. 2. This strategy works for potential functions of the type shown in panel (a) but obviously can fail for other potentials, for example, that shown in panel (b). The trajectory followed will depend on the "nesting" of the *effective catchment regions* of the smoothed potentials. The scale-dependent nesting patterns determine the system's *scaling properties*. Fig. 2a provides an example of *strong scaling*; Fig. 2b provides an example of *weak scaling*. These properties, which govern the likelihood of success of spatial averaging methods, will be discussed in more detail in Sec. 5.

2. Diffusion equation method (DEM). The DEM [11,12] provides a simple, direct method for variable-scale potential smoothing. It takes the potential $U(R)$ as the initial $t = 0$ boundary condition of the diffusion equation (where t is a formal "time" parameter distinct from physical time) and then uses mathematical "diffusion" over time-interval t to obtain the smoothed potential $U(R,t)$. That is, $U(R,t)$ is determined as the solution of

$$(2.1) \qquad \frac{\partial}{\partial t}U(R,t) = \nabla^2 U(R,t) \,,$$

with boundary condition

$$U(R,0) = U(R) \,.$$

The solution is obtained by convolution with a Gaussian kernel:

$$(2.2) \qquad U(R,t) = (4\pi t)^{-N/2} \int e^{-(R-R')^2/4t} \, U(R') \, dR' \,,$$

where N is the number of degrees-of-freedom. Thus, $\Lambda^2 = 2t\, I$ measures the spatial scale of the smoothing. It is a scalar multiple of the identity matrix I, so the smoothing is isotropic in the conformation space. $U(R,t)$ is an isotropically smoothed potential which can be used in the algorithm illustrated in Fig. 2.

DEM algorithm

1. Increase t to t_{hi} until $U(R, t_{hi})$ is convex or almost convex; then find a local minimum $R^0(t_{hi})$.
2. Track the position of $R^0(t)$, the local minimum of $U(R, t)$, as $t \to 0$. That is, alternately perform
   1. $t \to t - \Delta t$.
   2. Locally minimize $R^0 \ni \nabla U(R^0, t) = 0$.

The hope of the DEM is that this procedure will lead to the global minimum.

2.1. Tests of the DEM. The DEM has been applied [13] to the Lennard-Jones microcluster problem which consists of N atoms located in 3-space interacting with potential energy

$$(2.3) \qquad U(R) = U_0 \sum_{i<j}^{N_{atoms}} \left[\left(\frac{r_{ij}}{r_0} \right)^{-6} - 2 \left(\frac{r_{ij}}{r_0} \right)^{-12} \right],$$

where r_{ij} is the distance between atoms i and j and U_0 and r_0 are parameters with dimensions of energy and distance. Because $\lim_{r_{ij} \to 0} U(R) \to \infty$, the integral in (2.2) can diverge. The DEM practitioners cope with this problem by truncating the potential at a large, but arbitrarily chosen value [13]. This is believed to be acceptable because none of the r_{ij} are ever very small in the final solution.[2] (The need for this procedure comes from the fact that the DEM, while introducing a measure of spatial scale, does not introduce a measure of energy scale.) The truncated potential is then expanded in terms of a small number of Gaussians [13]. Since products of Gaussians can be analytically integrated, this reduces the evaluation of (2.2) to the algebra required to handle the sums. Using these methods Kostowicki et al. [13] tested the performance of the DEM on microclusters containing up to 55 atoms. Performance was mixed: global minima were identified in about two-thirds of the tested cases.

The DEM was next tested on clusters of up to eight water molecules [14]. Global minima were identified for $N \le 5$, but local, not global, minima were found for $N > 5$.

The DEM was also tested on the pentapeptide Met-enkephalin [15]. In this case all bond-lengths, bond-angles, and peptide-bond torsion-angles were held fixed and the remaining 19 torsion angles were allowed to vary. The energy was modeled using the ECEPP potential [16]:

$$(2.4) \qquad U = \sum_{i<j}^{N_{atoms}} \left[\frac{A_{ij}}{r_{ij}^{12}} + \frac{B_{ij}}{r_{ij}^{10}} + \frac{C_{ij}}{r_{ij}^{6}} + \frac{D_{ij}}{r_{ij}} \right] + \sum_{i}^{N_\theta} U_i^{tor}(\theta_i) .$$

---

[2] However, the choice of cutoff does affect the values of $U(R, t)$ in regions that are sampled during minimization. Thus, changing the cutoff may, in principle, affect the trajectory that is followed during minimization.

where the A_{ij}, B_{ij}, and C_{ij} are fixed parameters that depend on the specific ij^{th} atom-pair and the U_i^{tor} are terms that depend on the torsion angles θ_i. The r^{-12} and r^{-6} terms are used to model the Lennard-Jones potential, the r^{-10} term is part of a special hydrogen-bonding potential, and the r^{-1} models electrostatic interactions. The torsional terms reflect the proclivity of the angles to certain conformations but were excluded from the DEM analysis because they can not be easily modeled as sums of Gaussians. The DEM identified a local minimum in the vicinity of the global minimum but not the global minimum itself. The authors suggest that this was a consequence of inappropriate approximations that were made in performing convolution (1.6) in the presence of bond-length and bond-angle constraints, and report that the global minimum can be found by an improved implementation [17].

3. Adiabatic Gaussian density annealing (AGDA). Gaussian Density Annealing [18,19,20] begins by considering macromolecules as thermodynamic systems governed by the Gibbs/Boltzmann distribution $p_B(R)$. It assumes that p_B can be approximated by a single Gaussian which is the product of independent Gaussian distributions for the $i = 1 \ldots N$ individual atoms:

$$p_B(R) \approx p_{GDA}(R) \equiv \prod_{i=1}^{N} p_i(\mathbf{r}_i)$$

(3.1a)
$$\equiv \prod_{i=1}^{N} (2\pi\sigma_i^2)^{-\frac{3}{2}} e^{-\frac{1}{2}[\mathbf{r}_i - \mathbf{r}_i^0(\beta)]^2 / \sigma_i^2(\beta)}$$

(3.1b)
$$\equiv \det^{-\frac{1}{2}}(2\pi\Lambda^2) e^{-\frac{1}{2}[R - R^0(\beta)]\cdot\Lambda^{-2}(\beta)\cdot[R - R^0(\beta)]}$$

where $\mathbf{r}_i$ is the 3-vector coordinate of atom i and $\mathbf{r}_i^0$ and σ_i are 3-vector and scalar parameters that describe the individual atomic Gaussians. R^0 and Λ (determined by the $\mathbf{r}_i^0$ and σ_i, respectively) characterize the center and fluctuation tensor of the corresponding single multidimensional Gaussian. Because the fluctuations of the atoms are assumed to be independent and isotropic in the 3-space, Λ is a diagonal matrix. Both R^0 and Λ depend on the inverse temperature β. The dependence is determined by differentiating the first and second spatial moments of the Gibbs/Boltzmann distribution using the Gaussian approximation. Since

(3.2a)
$$R^0 = \langle R \rangle_{GDA}$$

(3.2b)
$$(\Lambda^2)_{ij} = \delta_{ij} \langle (R_i - R_i^0)^2 \rangle_{GDA},$$

we have

(3.3a) $\quad \dfrac{d}{d\beta} R^0 = \dfrac{d}{d\beta}\langle R\rangle_{GDA} \approx \dfrac{d}{d\beta}\langle R\rangle = -\langle U\,R\rangle + \langle U\rangle\,\langle R\rangle$

(3.3b) $\quad\qquad\qquad\qquad \approx -\langle U\,R\rangle_{GDA} + \langle U\rangle_{GDA}\,\langle R\rangle_{GDA}$

(3.3c) $\quad\qquad\qquad\qquad = -\Lambda^2\,\nabla_{R^0}\langle U\rangle_{GDA}\;,$

where

(3.4a) $\quad\qquad\qquad \langle f\rangle_{GDA} \equiv \displaystyle\int f(R)\,p_{GDA}(R)\,dR$

(3.4b) $\quad\qquad\qquad \langle f\rangle \equiv \displaystyle\int f(R)\,p_B(R)\,dR\;.$

Similarly we get

(3.5) $\quad\qquad\qquad \dfrac{d}{d\beta}(\Lambda^2)_{ii} \approx -(\Lambda^2)_{ii}^2\,\dfrac{\partial^2}{\partial(R_i^0)^2}\langle U\rangle_{GDA}\;.$

Eqs. (3.3c) and (3.5) determine the temperature evolution of R^0 and Λ in terms of the gradient and Hessian of $\langle U\rangle_{GDA}$ which is a Gaussian convolution of U:

(3.6a) $\langle U\rangle_{GDA} \equiv \langle U\rangle_{GDA}[R^0(\beta), \Lambda(\beta)]$

(3.6b) $\qquad = \det^{-\frac{1}{2}}[2\pi\Lambda^2(\beta)]\displaystyle\int U(R)\,e^{-\frac{1}{2}[R-R^0(\beta)]\cdot\Lambda^{-2}(\beta)\cdot[R-R^0(\beta)]}\,dR\;.$

Thus, $\langle U\rangle_{GDA}$ is a smoothed potential with Λ as the spatial averaging parameter. Eqs. (3.3c) and (3.5) give the

Gaussian density annealing algorithm

1. Start with initial conditions at sufficiently high temperature ($\beta_{lo} \approx 0$) where all conformations are almost equally likely and $\Lambda(\beta_{lo})$ is much greater than the interparticle separations. Randomly select large initial $\Lambda(\beta_{lo})$ and multiple random starting points $R^0(\beta_{lo})$.
2. For each starting point solve for $R^0(\beta)$ and $\Lambda(\beta)$ as $\beta \to \infty$ using Eqs. (3.3c) and (3.5).

Tsoo and Brooks [21] found that this procedure can be unstable because of the accumulation of errors resulting from the neglect of anharmonicities in approximation (3.1a). They showed that performance was improved by occasionally allowing R^0 to find its stationary point

(3.7) $\qquad\qquad \nabla_R\,\langle U\rangle_{GDA}(R, \Lambda)|_{R=R^0} = 0$

before proceeding with evolution in β. This led to the *Adiabatic Gaussian Density Annealing* (AGDA) method [20]:

AGDA algorithm

1. Start with initial conditions at sufficiently high temperature ($\beta_{lo} \approx 0$) where all conformations are almost equally likely and $\Lambda(\beta_{lo})$ is much greater than the interparticle separations. Randomly select a large initial $\Lambda(\beta_{lo})$ and find a local minimum $R^0(\beta_{lo})$ of $\langle U \rangle_{GDA}(R^0, \Lambda)$.

2. Track $R^0(\beta)$ and $\Lambda(\beta)$ as $\beta \to \infty$ by alternately performing
   1. $T \to T - \delta T$
   2. Locally minimize $R^0 \ni \nabla_{R^0} \langle U \rangle_{GDA}(R^0, \Lambda) = 0$
   3. Evolve Λ according to Eq. (3.5).

This algorithm is similar to the DEM algorithm except that Λ is determined as a function of β by (3.5) and the smoothed potential is obtained by averaging with a Gaussian having limited anisotropy. Both methods seek the global minimum by tracing a single trajectory obtained by performing a sequence of local minimizations of a Gaussian convolution of U. We can regard the DEM as a special case of the AGDA with Λ restricted to be a multiple of the identity matrix, $\Lambda = \lambda I$. In this case λ will be a monotonically-decreasing function of β and can be regarded as the independent parameter that replaces the DEM t. The DEM and AGDA smoothed potentials are related by $U(R, t) = \langle U \rangle_{GDA}(R, \sqrt{2t}I)$.

The AGDA has been tested on a variety of problems including Lennard-Jones microclusters, water clusters, a covalently-linked Lennard-Jones homopolyer, and model proteins composed of hydrophobic, hydrophilic and neutral "amino acids" [20,21]. Its performance was better than that of simulated annealing using molecular dynamics sampling and was comparable with that of the DEM: the global minimum was found in many but not in all cases. Tsoo and Brooks [21] concluded that the difficulties could primarily be associated with two problems: (1) errors in the solution of $d\Lambda^2/d\beta$ accumulate with increasing β, and (2) the single Gaussian (3.1b), while satisfactory at high and low temperatures, is inaccurate at intermediate temperatures.

4. Packet annealing method (PAM). The PAM, like AGDA, performs global minimization by analyzing a coarse-grained approximation to the Gibbs/Boltzmann distribution [22,23,24].[3] However, it differs from AGDA in a number of ways, particularly in its use of a multi-Gaussian approximation to p_B. The conceptual motivation of the PAM is illustrated in Fig. 3 which displays a 2-dimensional potential (panel a) and the corresponding Gibbs/Boltzmann distributions at different temperatures (panel b). It is evident that p_B can be roughly decomposed into regions of sep-

---

[3] Some minor changes in notation are made here relative to Refs. [22] and [23]. In particular, the definitions of Λ^2 differ by a factor of 2.

arated probability density, but the number of regions needed varies with temperature. Within each region α, p_B can be approximated by a Gaussian packet parameterized by an average position $R_\alpha^0(\beta)$ and fluctuation tensor $\Lambda_\alpha(\beta)$ (panel c). In contrast with AGDA, the PAM Gaussians are allowed to be fully anisotropic; that is, the $\Lambda_\alpha(\beta)$ can be general positive semi-definite[4] symmetric matrices, not necessarily diagonal matrices. This permits the principal axes of fluctuation to lie in any directions in conformation space. This freedom is important for large molecules where fluctuations may involve correlated motions of many atoms.

Mathematically, the PAM multi-packet approximation to p_B is

$$(4.1) \qquad p_B^{\frac{1}{2}}(R) \approx \widetilde{p}^{\frac{1}{2}}(R) \equiv \sum_{\{\alpha\}} \phi_\alpha^0(V_\alpha^0, R_\alpha^0, \Lambda_\alpha^0; R)$$

where the dependence of all terms on β is implicit. The $\phi_\alpha^0(V_\alpha^0, R_\alpha^0, \Lambda_\alpha^0; R)$ are the *characteristic packets*:

$$(4.2) \qquad \phi_\alpha^0 \equiv e^{-\frac{1}{2}[\beta V_\alpha^0 + \frac{1}{2}(R-R_\alpha^0)\cdot\Lambda_\alpha^{-2}\cdot(R-R_\alpha^0)]} .$$

Free-energies and other thermodynamic parameters can be calculated for each packet.

4.1. Packet equations. The packets are intended to correspond to the metastable macrostates described in Sec. 1.1. The fundamental characteristic of a metastable macrostate is that the time scale for internal equilibration of probability density within the macrostate, τ_α^{local}, be much smaller than the time scale for transitions out of the macrostate, τ_α^{global}. We have shown that such regions can be identified by looking for characteristic packets whose first and second spatial moments are stable to first-order in time [25]. That is, we require

$$(4.3a) \qquad \frac{\partial}{\partial t}\langle R\rangle_\alpha^{PAM}|_{t=0} \equiv 0$$

$$(4.3b) \qquad \frac{\partial}{\partial t}\langle (R-R_\alpha^0)_i(R-R_\alpha^0)_j\rangle_\alpha^{PAM}|_{t=0} \equiv 0$$

where

$$(4.4) \qquad \langle f\rangle_\alpha^{PAM} \equiv \frac{\int f p_B^{\frac{1}{2}}\phi_\alpha^0\, dR}{\int p_B^{\frac{1}{2}}\phi_\alpha^0\, dR}.$$

Using the Smoluchowski equation, which describes diffusion in the presence of an external potential, we have shown [25] that this implies the

---

[4] Zero eigenvalues will be associated with continuous symmetries of $U(R)$ such as translation and rotation. See Ref. [23].

Packet Equations

Integral form	**Differential form**

$$(4.5a) \qquad R_\alpha^0 = \langle R \rangle_\alpha^{PAM} \qquad\qquad \left. \frac{\partial \widetilde{H}_{\Lambda_\alpha,T}(R)}{\partial R} \right|_{R=R_\alpha^0} = 0$$

$$(4.5b) \quad \Lambda_\alpha^2 = \langle (R - R_\alpha^0)(R - R_\alpha^0) \rangle_\alpha^{PAM} \qquad \left. \frac{\partial^2 \widetilde{H}_{\Lambda_\alpha,T}(R)}{\partial R^2} \right|_{R=R_\alpha^0} = \Lambda_\alpha^{-2}/2\beta$$

where

$$(4.6) \qquad \widetilde{H}_{\Lambda,T}(R) \equiv -2\beta^{-1} \log \left[\det^{-\frac{1}{4}} (2\pi\Lambda^2) \right. \\ \left. \times \int e^{-\frac{1}{2}\beta U(R')} e^{-\frac{1}{4}(R-R')\cdot\Lambda^{-2}\cdot(R-R')} \, dR' \right]$$

is the *effective energy*. The close relationship of (4.6) to the macrostate occupation probability (1.3) is evident.[5] It can be shown that $\widetilde{H}_{\Lambda_\alpha,T}(R_\alpha^0)$ approximates the free-energy of macrostate α and that Eqs. (4.5) identify the local minima of the free-energy of characteristic packet α in the $\{R^0, \Lambda\}$ parameter space [23,25]. These equations determine the characteristic packet parameters R_α^0 and Λ_α.

The integral forms of the packet equations are used for computation and are solved iteratively. The differential forms illuminate the relationship between the PAM and the DEM and AGDA. $\widetilde{H}_{\Lambda,T}(R)$ plays the role of $U(R,t)$ in the DEM and $\langle U \rangle_{GDA}$ in the AGDA. However, $\widetilde{H}_{\Lambda,T}$ is the Gaussian convolution of $p_B^{\frac{1}{2}}$ rather than the Gaussian convolution of U, as in the DEM and AGDA. The first-order packet equation (4.5a) is analogous to (3.7) of the DEM and AGDA. The second-order equation (4.5b) is analogous to (3.5) of the AGDA in that both equations determine the size of a Gaussian packet. However, in contrast with the explicit AGDA equations, Eqs. (4.5) are a self-consistent coupled set of implicit equations. Thus, their solution trajectories $\{R_\alpha^0(\beta), \Lambda_\alpha(\beta)\}$ can contain bifurcations and other singularities, and the number of solutions to Eqs. (4.5) can vary with temperature. These packet *branching transitions* occur when the system temperature is lowered to the point where an energy barrier separating two regions of configuration space can sufficiently retard the exchange of probability density. Fig. 3 provides examples of how the number of packets can vary with temperature: packet α, which is unique at high temperatures, branches into *child* packets β and γ at its *transition temperature*.

---

[5] The integrand of (4.6) is the geometric mean of p_B and its characteristic packet approximation.

Packet γ subsequently branches into states δ and ϵ at a lower temperature. A simple analytic example of packet branching is described in Ref. [22]. Numerically computed examples of packet branching in more complicated physical systems and the hysteresis phenomena associated with branching transitions are described in Ref. [23].

4.2. Relationship between the PAM, AGDA and DEM minimization equations. If U is bounded, then in the high-temperature limit $\beta \to 0$ the $\exp(-\frac{\beta}{2}U)$ term in the integrand of $\widetilde{H}_{\Lambda,T}$ can be expanded in a Taylor's series to give

High temperature limit

(4.7a)
$$\lim_{\beta \to 0} \widetilde{H}_{\Lambda,T}(R) = \det^{-\frac{1}{2}}(4\pi\Lambda^2) \int U(R')\, e^{-\frac{1}{4}(R-R')\cdot\Lambda^{-2}\cdot(R-R')}\, dR'$$

$$- \beta^{-1} \log[\det^{\frac{1}{2}}(8\pi\Lambda^2)]$$

(4.7b)
$$\equiv \widetilde{U}_{\Lambda,\infty}(R) - \beta^{-1} \log[\det^{\frac{1}{2}}(8\pi\Lambda^2)]$$

The second line has been parsed to have the same form as expression (1.5) for the free-energy: the first term is the high-temperature limit of the mean-energy and the second (R-independent) term is an entropic contribution proportional to the logarithm of the conformation space volume. If Λ is restricted to be diagonal then

(4.8)
$$\widetilde{U}_{\Lambda,\infty} = \langle U \rangle_{GDA}(R, \sqrt{2}\Lambda).$$

That is, up to a difference in the definition of Λ, the PAM effective energy reduces to the AGDA smoothed energy in high-temperature limit. If Λ is further restricted to be a multiple of the identity matrix then

(4.9)
$$\widetilde{U}_{\Lambda,\infty} = U(R, \Lambda^2) .$$

That is, the PAM effective energy has the same form as the DEM smoothed energy.

Eqs. (4.7) is only valid when $U(R)$ is bounded. For singular $U(R)$ appearing in macromolecular energy functions like (2.4), replacement of $\exp(-\frac{\beta}{2}U)$ by its Taylor's series expansion is invalid and the integral in (4.7) does not converge, even though $\widetilde{H}_{\Lambda,T}$ is still well-defined for finite temperature. However, as discussed for the DEM and AGDA, (4.7) can be salvaged by truncating $U(R)$ to make it bounded.

Using (4.7–4.9) we see that the first-order packet equation (4.5a) is identical to the DEM and AGDA first-order equations in the high-temperature limit. The second-order packet equation implies that $\Lambda^2 \to \infty$

as $\beta \to 0$, but it can also be used to compute $d\Lambda^2/d\beta$. Multiplying (4.5b) by β and differentiating yields

(4.10)
$$\frac{d\Lambda^{-2}}{d\beta} = 2 \left.\frac{\partial^2 \widetilde{U}_{\Lambda,\infty}(R)}{\partial R^2}\right|_{R=R^0}.$$

If we restrict Λ to be a diagonal matrix (as in the AGDA), this can be rewritten as

(4.11)
$$\frac{d\Lambda_{ii}^2}{d\beta} = -2(\Lambda_{ii}^2)^2 \left.\frac{\partial^2 \widetilde{U}_{\Lambda,\infty}(R)}{\partial R^2}\right|_{R=R^0}.$$

Eq. (4.11) [in combination with (4.7)] is exactly equivalent to the AGDA second-order equation (3.5). We conclude that the AGDA can be regarded as a special case of the PAM with the following restrictions:

1. Λ is restricted to be a diagonal matrix.
2. Only one packet is considered.
3. Analysis is restricted to the high temperature limit $\beta \approx 0$.

As previously discussed, the DEM can be obtained by further restricting Λ to be a multiple of the identity matrix.

Eqs. (4.7-4.9) show that the DEM and AGDA smoothed potentials are high-temperature approximations to the PAM effective energy. However, these linear (in U) approximations become invalid as $T \to 0$. Thus, from the standpoint of the PAM, it is necessary to use to complete, non-linear expression for $\widetilde{H}_{\Lambda,T}$ (4.6), not just the linear approximation, for annealing.

4.3. Computation of $\widetilde{H}_{\Lambda,T}$. $\widetilde{H}_{\Lambda,T}$, in contrast to the DEM $U(R,t)$ and the AGDA $\langle U \rangle_{GDA}$, is well-defined even for singular potentials like Lennard-Jones, and no *ad hoc* potential truncation needs to be used. However, it is more complicated to evaluate: the non-linearity in Eq. (4.6) means that it can not be reduced to a simple Gaussian integral, even if $U(R)$ is approximated as a sum of Gaussians. Instead, we have proposed [22] the use of an "effective harmonic approximation" that exploits the *partial separability* of macromolecular functions. Partial separability means that U can be written as a sum of terms, each depending only on a few variables:

(4.12)
$$U(R) = \sum u_i(\xi_i),$$

where the ξ_i are variables such as interpair distances or torsion angles. The ECEPP potential [16] is an example of a potential having this form. We approximate

(4.13)
$$\widetilde{H}_{\Lambda,T}(R) = \sum_i \widetilde{h}_{\lambda_i,T}(\xi_i),$$

where $\widetilde{h}_{\lambda_i,T}$ is evaluated from a reduced form of Eq. (4.6) that only involves the ξ_i. The λ_i are determined by a self-consistent procedure [26].

Because there are only a few different types of u_i (e.g., representing Van derWaals and electrostatic interactions), each depending on a small number of parameters (i.e., T, λ_i, and ξ_i), the effective-energy integrals can be pre-calculated and tabulated in look-up tables. Thus the computational expense of evaluating $\widetilde{H}_{\Lambda,T}$ during run-time is minimal.

This approach has been tested in single-packet annealing of the Lennard-Jones microcluster problem with ≤ 24 atoms [22]. The full packet annealing procedure was not used; instead $\widetilde{H}_{\Lambda,T}$ was calculated at fixed T and Λ was restricted to be isotropic. As expected (since this differed from the DEM only in the specific manner by which the smoothed potential was calculated), performance was similar to that of the DEM: the global minimum was found in about three-fourths of the tested cases. A build-up method based on this restricted procedure, when applied to larger microclusters discovered a new global minimum for the 72 atom microcluster which had not been discovered by other procedures [27]. Furthermore, comparative tests showed that the use of $\widetilde{H}_{\Lambda,T}$ in simulated annealing, even with isotropic Λ, enhanced efficiency about 10-fold relative to standard simulated annealing using $U(R)$ for microclusters containing 16 or 27 atoms [28]. We believe performance will be improved by use of anisotropic Λ.

4.4. Trajectory diagrams. At high temperature the packet equations will have at most only a small number of characteristic packet solutions (e.g., as in the T_{hi} panels of Figs. 2 and 3). As temperature is decreased, each solution trajectory $\{R^0(\beta), \Lambda(\beta)\}$ can be traced down to its transition temperature and the child packets identified. This procedure can be continued recursively as temperature is reduced. In principle, a complete *trajectory diagram* which displays the hierarchical structure of the entire energy landscape can be constructed. A simple example of a R^0 vs T trajectory diagram corresponding to a one-dimensional potential is shown in Fig. 4. Trajectory diagrams can be plotted for every relevant macrostate variable. The thermodynamic properties discussed in Sec. 1.1 are of particular importance. For example, Fig. 3e displays part of the free-energy vs T trajectory diagram for the two-dimensional example we have discussed. Complete trajectory diagrams for Lennard-Jones microclusters with 6, 7 and 8 atoms are presented in Ref. [23].

4.5. Macromolecular trajectory diagrams. Because of their covalent linkages, macromolecular motions are highly non-linear when expressed in Cartesian coordinates. Thus, we can not expect even an anisotropic Gaussian model to accurately describe fluctuations in these coordinates. Instead, we describe these systems using a dual representation: torsion-angle variables are used to describe individual and average conformations (e.g., R_α^0) while inter-atomic distance variables are used to characterize fluctuations (i.e., Λ_α) and macrostate boundaries. In the distance variable formulation, packet bifurcations correspond to gaps in the distance probability distributions. The appearance of a probability gap, as in Fig. 5, indicates the presence of an *effective energy barrier* in the projection of the probability density into the distance variable d_i. This barrier splits the probability distribution into two components corresponding to packet bifurcation.

The distance variable representation is highly redundant: there are $N(N-1)/2$ distances between N atoms, so most of the distances are highly correlated. Accordingly, effective energy barriers appear almost simultaneously in multiple distance variables. Thus, only a subset of distance variables needs to be monitored for packet branching. We expect that the analysis will not be sensitive to the particular selection of this subset of distance variables as long as it is adequate to determine the conformation of the macromolecule. Since the distance variables are already being calculated to evaluate the potential U, the analysis involves little additional overhead. It is important to note that the torsion-angle coordinates do not need to be determined from the distances since torsion-angles are simultaneously being tracked in the dual representation.

This approach was used to calculate the free-energy trajectory diagram for Met-enkephalin. At very high temperatures Met-enkephalin has only one thermodynamically stable macrostate which fluctuates throughout all of conformation space. The first branching transition occurs at ~ 10 kcal/mole where an effective energy barrier in the distance variables governing the orientation of the carboxyl-terminal methionine causes the single high-temperature macrostate two split into substates, one one of which is shown in Fig. 6. Effective energy barriers which appear in other distances at lower transition temperatures cause further bifurcations and the appearance of more low-temperature metastable macrostates. The proliferation of macrostates continues until, as $T \to 0$, each local minimum supports a macrostate. Because of the large number of local minima in the Met-enkephalin energy landscape [estimated at $\sim O(10^9)$], only a subset of the trajectory diagram can be calculated. The trajectory which leads from the unique high-temperature macrostate to the global minimum [16,29] and a sampling of other trajectories are displayed in Fig. 7.

This diagram contains a wealth of coarse-grained information about the landscape. Each trajectory point represents a macrostate parameterized by a mean conformation R^0 and fluctuation tensor Λ. The hierarchical relationships between the macrostates is embedded in the topological structure of the trajectory diagram. It is important to note that the trajectory which leads to the global minimum (closed boxes) passes through a temperature region $(-0.2 \leq \log(T) \leq 0.4)$ where its free-energy is very high (i.e., probability is very low). This "weak scaling" behavior in free-energy will be discussed in Sec. 5.

4.6. Global minimization using the full PAM. While partial implementations of the PAM strategy for global minimization have been implemented and tested (Sec. 4.3), a complete implementation, using all the PAM tools we have discussed, has not yet been developed. PAM global minimization will involve following multiple packet trajectories as temperature is lowered. As seen for Met-enkephalin, it will not be possible to track all trajectories and success will depend on our ability to make branch selections that lead to the global minimum. If only one path were to be followed a reasonable choice might be to follow the child that has the lowest free-energy. This procedure will succeed for simple problems like those illustrated in Figs. 2a, 3 and 4. However, it will fail for problems like that illustrated in Fig. 2 and for Met-enkephalin (Fig. 7) where the path leading to the global minimum passes through a low probability (high free-energy) region at intermediate temperatures. Other selection procedures can also be considered, such as selecting those children having the lowest mean-energy. A more robust procedure would be to follow multiple trajectories, including some with low free-energy and some with low mean-energy. This strategy clearly lends itself to coarse-grained parallel computing. The critical task is to develop appropriate resource-allocation strategies. We hope to be guided in this by studying trajectory diagrams from related problems to identify common properties of families of related problems (e.g., protein energy landscapes). Then, it will be possible to implement a complete [22]

PAM algorithm

1. Begin at high temperature T_{hi} where the small number of solutions to the packet equations (4.5) can be found. Calculate $\{R_\alpha^0(T_{hi}), \Lambda_\alpha(T_{hi})\}$ for each solution α.

2. Track the $\{R_\alpha^0(T), \Lambda_\alpha(T)\}$ as $T \to 0$, allowing for packet branching to child packets. That is, alternately perform

   1. $T \to T - \Delta T$.
   2. For each packet α, perturbatively update the packet solutions $\{R_\alpha^0(T), \Lambda(T)\}$ using (4.5).
   3. Test each packet for branching into child packets.
   4. Allocate resources: determine which packets are to be followed.

5. Scaling properties. The ability of variable-scale annealing methods to find global minima will depend on the structure of the energy landscape under consideration. It is commonly suggested that macromolecular landscapes have "funneling" properties that guide ensembles of systems to global minima as temperature is lowered [30,31]. Fig. 2 provides an example of a landscape with this property. It would be interesting to know if energy landscapes (like those of macromolecules) that are generated by the summation of large numbers of quasi-independent terms [as in Eq. (4.12)] statistically tend to have this property. Whether or not this is the case, biological macromolecules may have evolved funneling properties for efficient folding. However, this is not guaranteed, and computable tests to determine whether real proteins have this property or not are needed.

To clarify the discussion, we characterize landscapes according to their *scaling properties*. We say that a landscape displays *strong scaling* if the position of its global minimum is relatively invariant (i.e., invariant up to the size scale under consideration) when the potential is "viewed" (i.e., smoothed) at different temperatures and/or spatial scales. In this case the catchment regions of the temperature-dependent effective potentials are "nested" so that ensemble members are funneled towards the global minimum as in Fig. 2a. When this is not the case, we say that a landscape displays *weak scaling*. This can occur when the global minimum lies in a narrow well that is outside the region spanned by the large spatial scale catchment regions that dominate at high temperature as in Fig. 2b.

Scaling properties can be identified by graphical inspection of one- and two-dimensional potentials, but other methods are needed to analyze high-dimensionality potentials. This is provided by the trajectory diagrams. The lower panels of Fig. 2 schematically display the free-energy trajectory diagrams corresponding to the strong and weak scaling potentials. In the strong free-energy scaling case, the global minimum at $T = 0$ is reached by the path that begins with the single high-temperature trajectory and which chooses the state of lowest free-energy at every branch point. This is not so in weak free-energy scaling where the trajectory which leads to the global minimum has high free-energy at high temperature. This occurs because of its small size and low entropy[6] The associated "trajectory crossing" behavior is a sign of weak scaling. This analysis can be performed even for high-dimensionality problems. For example, the global minimum of the Met-enkephalin energy landscape lies within a macrostate which has high free-energy (low probability) at intermediate temperatures (see Fig. 7). We conclude that Met-enkephalin has weak free-energy scaling.

We expect that landscape scaling properties will play an important role in determining the difficulty of global minimization; it will be much easier in the case of strong scaling or when the scaling is not "too" weak. Multi-trajectory methods may be needed when scaling is very weak. The degree of weakness of the scaling will determine how many of the low-lying trajectories will have to be traced to find a path to the global minimum. Although we have emphasized free-energy scaling properties, scaling can be defined in other variables as well. For example, while Fig. 7 shows that Met-enkephalin has very weak free-energy scaling, analysis of its mean-energy trajectory diagram shows that it has fairly strong mean-energy scaling [32]. This means that the Met-enkephalin global minimum can be easily found by following the low-lying mean-energy trajectories. It will be important to use trajectory diagram analysis to see if any statements can be made about the free-energy and mean-energy scaling properties of larger proteins

---

[6] From Eq. (1.5) we see that entropy dominates free-energy at high temperature.

and proteins in general. As discussed in Sec. 4.6, this information can be important for algorithm design.

6. Conclusion. The general characteristics of the DEM, AGDA and PAM are compared in Table 1. Except for the selection of initial conditions, all three are deterministic methods. As discussed in Secs. 3 and 4.2, we can view the DEM as a special case of AGDA and AGDA as a special case of the PAM.

The DEM is the simplest and most extensively tested and will probably be the method of choice if it is capable of solving more complicated problems. But it is not clear that the severe simplifications that it employs, particularly the use of isotropic averaging and the restriction to a single Gaussian packet, will give it sufficient versatility to solve large proteins. AGDA provides the increased power of partially anisotropic averaging at the cost of more computational expense. To date, no dramatic differences between the performance of the DEM and AGDA have been observed [20], but we believe that the benefits of anisotropic averaging will become more pronounced as the complexity of the systems under investigation increases. Within the assumption that the Gibbs/Boltzmann probability distribution can be modeled by a single Gaussian, it provides a model for physical annealing of a thermal system. However, this assumption may ignore important anharmonic properties of the landscape in the intermediate temperature range [21].

The PAM is the most versatile, but also the most complicated, of these methods. In principle, it can directly model the diffusive evolution of thermal systems and accommodate highly anharmonic potentials. While the DEM and AGDA minimize macrostate mean energies, the PAM can minimize macrostate free-energies. As discussed in Sec. 1.1, this is physically attractive since the free-energies are directly related to the macrostate occupation probabilities. The most significant difference between the PAM and the other methods is that its global minimization search trajectories bifurcate as temperature is lowered and algorithms for distributing computational resources will play an important role. The hierarchical multiplicity of search trajectories that is generated is a deterministic replacement for the random multiplicity of trajectories generated by stochastic methods. The resultant trajectory diagrams are valuable in themselves; they provide a new tool for characterizing the scaling and hierarchical properties of energy landscapes and "road-maps" for describing coarse-grained pathways through conformation space. However, we do not yet know if the the complexity of the PAM as a global minimization tool will be merited by improved performance since its individual components have only been tested separately, not in a full parallel implementation.

Variable-scale coarse-graining methods may be useful even when the global minimum is not found. It is not known whether all proteins attain thermodynamic equilibrium and fold to the global minimum of free-energy in all cases or whether some are kinetically-trapped in local minima. To the extent that the variable-scale coarse-graining methods mirror the physical folding process, even local minima that they identify may correspond to physically relevant states.

In summary, the DEM, AGDA and PAM provide examples of ways by which continuously-variable coarse-graining can be applied to global minimization of macromolecular energy functions. The field is in an exciting nascent state and the methodologies are rapidly evolving. While interesting results have been obtained using small model test systems, it remains to be seen if these methods will work on practical problems of larger size.

Acknowledgments. Many of the PAM ideas discussed here were developed in collaboration with Bruce Church, Jason Gans, Matej Orešič, and Alex Ulitsky. I thank them for many invigorating discussions. In particular, I thank Bruce Church for use of his unpublished data and for preparation of many of the figures. This work was supported by NIH grant GM48874 and AFOSR grant F49620.

TABLES

Comparison of DEM, AGDA and PAM

TABLE 1

*Characteristics of the Diffusion Equation Method (DEM), Adiabatic Gaussian Density Annealing (AGDA), and the Packet Annealing Method (PAM). *Multiple randomly-selected starting conformations may be used; one trajectory is traced from each.*

	DEM	AGDA	PAM
Gaussian convolution of	$U(R)$	$U(R)$	$p_B(R)$
Transformation of $U(R)$	Linear	Linear	Non-linear
Temperature-dependence	No	Yes	Yes
Minimization variable	Energy	Energy	Free-energy
Spatial scale parameter	Scalar	Diagonal matrix	Symmetric matrix
	isotropic	anisotropic	anisotropic
		fixed axes	variable axes
Search trajectory	Single*	Single*	Hierarchical family

FIGURES

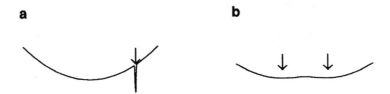

FIG. 1. *Physically irrelevant mathematical minima.*

Mathematical minima will be physically unimportant if they are too narrow (a) or two shallow (b) relative to the relevant spatial or energy scales.

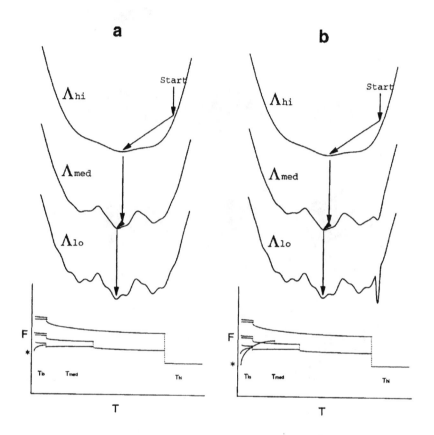

FIG. 2. *Progressive search by variable-scale global minimization methods.*

Each upper panel shows a sequence of smoothed objective functions obtained by convolution with Gaussians having widths $\Lambda_{hi} > \Lambda_{med} > \Lambda_{lo}$ according to Eq. (1.6). The basic search procedure is illustrated by the arrows. An initial search, beginning at Start *identifies the single minimum of $f(R, \Lambda_{hi})$. Using this as a next starting point, the subsequent search identifies the minimum of the associated catchment region of $f(R, \Lambda_{med})$. This procedure is repeated as Λ is decreased. The lower panels roughly indicate the types of free-energy vs temperature trajectory diagrams that correspond to these potentials. See Ref. [23] for accurately computed trajectory diagrams. Panels: (a) A "strong scaling" case where this strategy finds the global minimum. (b) A "weak scaling" case where a single sequence of searches using this strategy does not find the global minimum.*

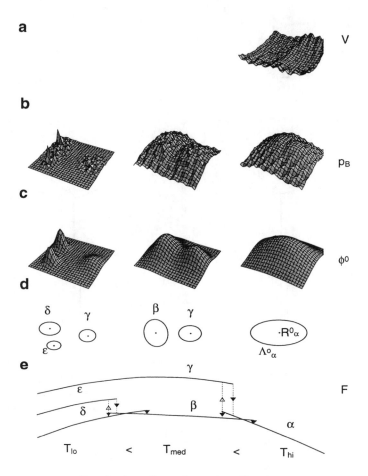

FIG. 3. *Annealing using metastable states.*

*(a)   A   model   two-dimensional   potential $U(r_1, r_2)$.   (b)   The   corresponding
Gibbs/Boltzmann probability distribution p_B at three temperatures, $T_{hi} > T_{med} > T_{lo}$.
(c) Superposition of the squared characteristic packets $(\phi_\alpha^0)^2$ that are solutions to the
packet equations at the three temperatures. (A large number of characteristic packets,
corresponding to the very small-scale fluctuations of U, will appear at lower temperatures.) (d) The characteristic packets are characterized by the positions of their center-
of-masses, R^0, and by their mean-square fluctuation tensors Λ, represented here by el-
lipses. (e) Free-energy vs temperature trajectory diagram for this temperature range.
Solid lines represent metastable state trajectories and dotted lines represent transitions.
The discontinuities in the trajectories correspond to branch points at which packets bifurcate. Solid arrowheads indicate "escape" or preferred "capture" transitions. Open
arrowheads indicate unpreferred transitions (see [23] for details). (Reprinted from [24].)*

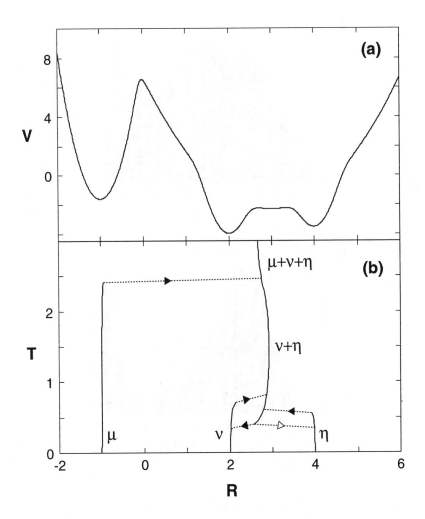

FIG. 4. *One-dimensional potential and corresponding free-energy trajectory diagram.*

The potential is shown in panel (a). The R^0 vs T diagram is shown in panel (b). Transitions are identified as described in Fig. 3. (Adapted from Ref. [23].)

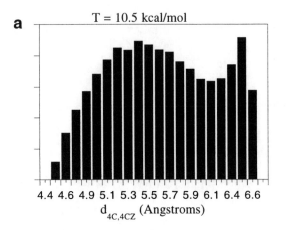

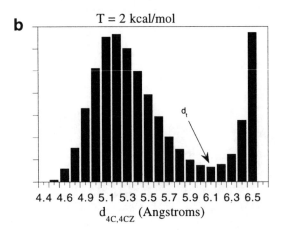

FIG. 5. *Projected probability distribution between two atoms in Met-enkephalin at high temperature (a) and at low temperature (b).*

At $T = 3.6$ *kcal/mol the distribution satisfies the criteria for bifurcation into two metastable states: one with* $d < d_t$ *and one with* $d > d_t$. *(B. W. Church, unpublished data.)*

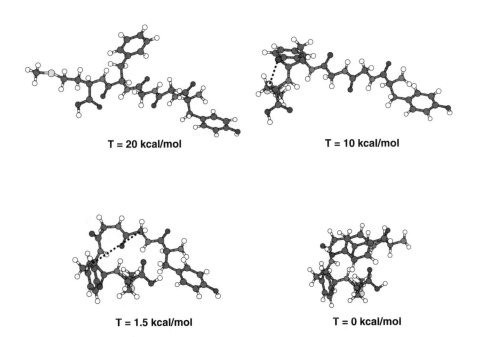

T = 20 kcal/mol   T = 10 kcal/mol

T = 1.5 kcal/mol   T = 0 kcal/mol

FIG. 6. *Examples of macrostates appearing during annealing of Met-enkephalin.*

The manner in which conformational structure progressively appears is displayed for a few of the packet bifurcations that occur during annealing of Met-enkephalin from T = 20 kcal/mole to 0 kcal/mole. The dashed lines indicate distance variables that identify transitions for single to bifurcated packets. The regions shown in an extended conformation are fluctuating without constraint. (B. W. Church, unpublished data.)

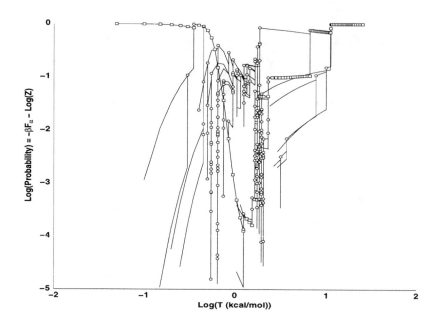

FIG. 7. *Met-enkephalin free-energy/probability trajectory diagram.*

The free-energies are scaled and shifted so that the vertical axis corresponds to the log(probability) for each state. The trajectory which leads from the unique high-temperature macrostate to the Met-enkephalin global minimum is identified by open boxes. As $T \to 0$ this macrostate captures all the probability. Other trajectories are indicated by open circles. The trajectory diagram has been pruned and many low-probability trajectories are terminated. The gap between the global minimum trajectory and the high- probability trajectories in the region $\log(T) \approx 0.2$ only reflects pruning, not an absence of trajectories in this region. (B. W. Church, unpublished data.)

REFERENCES

[1] H.S. CHAN AND K.A. DILL, Phys. Today **46**, 24 (1993).

[2] C.E. BUGG, W.M. CARSON, AND J.A. MONTGOMERY, Sci. Am. **269**, 92 (1993).

[3] W.C. GUIDA, Curr. Biol. **4**, 777 (1994).

[4] A.T. BRUNGER, J. KURIYAN, AND M. KARPLUS, Science **235**, 458 (1987).

[5] T.L. BLUNDELL, B.L. SIBANDA, M.J.E. STERNBERG, AND J.M. THORNTON, Nature **326**, 347 (1987).

[6] T. BLUNDELL, et al., Eur. J. Biochem. **172**, 513 (1988).

[7] H.A. SCHERAGA, in *Reviews in Computational Chemistry*, Vol. 3, edited by K.B. LIPKOWITZ AND D.B. BOYD (VCH Publishers, New York), pp. 73–142 (1992).

[8] L.D. LANDAU AND E.M. LIFSHITZ, *Statistical Physics* (Addison–Wesle y, Reading, Mass., 1969).

[9] Z. WU, SIAM J. Optimization **6**, (1996, in press).

[10] T. COLEMAN AND Z. WU, J. Global Optimization **8**, 49 (1996).

[11] L. PIELA, J. KOSTROWICKI, AND H.A. SCHERAGA, J. Phys. Chem. **93**, 3 339 (1989).

[12] J. KOSTROWICKI AND H.A. SCHERAGA, in *Global Minimization of Noncon vex Energy Functions: Molecular Conformation and Protein Folding: DIMACS Workshop, March 20–21, 1995*, Vol. 23 of *DIMACS Series in Discrete Mathematics and Theoretical Computer Science*, edited by P. PARDALOS, D. SHALLOWAY, AND G. XUE (American Mathematical Society, Providence, RI, 1996) , p. 123.

[13] J. KOSTROWICKI, L. PIELA, B.J. CHERAYIL, AND H.A. SCHERAGA, J. Phys. Che m. **95**, 4113 (1991).

[14] R.J. WAWAK, M.M. WIMMER, AND H.A. SCHERAGA, J. Phys. Chem. **96**, 51 38 (1992).

[15] J. KOSTROWICKI AND H.A. SCHERAGA, J. Phys. Chem. **96**, 7442 (1992) .

[16] G. NÉMETHY, et al., J. Phys. Chem. **96**, 6472 (1992).

[17] J. KOSTROWICKI AND H.A. SCHERAGA, personal communication.

[18] J. MA AND J.E. STRAUB, J. Chem. Phys. **101**, 533 (1994).

[19] P. AMARA, J. MA, AND J.E. STRAUB, in *Global Minimization of Noncon vex Energy Functions: Molecular Conformation and Protein Folding: DIMACS Workshop, March 20–21, 1995*, Vol. 23 of *DIMACS Series in Discrete Mathematics and Theoretical Computer Science*, edited by P. PARDALOS, D. SHALLOWAY, AND G. XUE (American Mathematical Society, Providence, RI, 1996) , p. 1.

[20] J.E. STRAUB, in *New Developments in Theoretical Studies of Proteins* , edited by R. ELBER (World Scientific, Singapore, 1996) (in press).

[21] C. TSOO AND C.L. BROOKS III, J. Chem. Phys. **101**, 6405 (1994).

[22] D. SHALLOWAY, J. Global Optimization **2**, 281 (1992).

[23] M. OREŠIČ AND D. SHALLOWAY, J. Chem. Phys. **101**, 9844 (19 94).

[24] B.W. CHURCH, M. OREŠIČ, AND D. SHALLOWAY, in *Global Minimization of Nonconvex Energy Functions: Molecular Conformation and Protein Folding: DIMACS Workshop, March 20–21, 1995*, Vol. 23 of *DIMACS Series in Discrete Mathematics and Theoretical Computer Science*, edited by P. PARDALOS, D. SHALLOWAY, AND G. XUE (American Mathematical Society, Providence, RI, 1996), p. 41.

[25] D. SHALLOWAY, J. Chem. Phys. **105**, 9986 (1996).

[26] D. SHALLOWAY, unpublished results.

[27] T. COLEMAN, D. SHALLOWAY, AND Z. WU, J. Global Optimization **4**, 17 1 (1994).

[28] T. COLEMAN, D. SHALLOWAY, AND Z. WU, Comput. Optim. Appl. **2**, 145 (1993).

[29] Z. LI AND H.A. SCHERAGA, Proc. Natl. Acad. Sci. USA **84**, 6611 (1987).

[30] K.S. DILL, Curr. Opin. Struct. Biol. **3**, 99 (1993).

[31] J.D. BRYNGELSON, J.N. ONUCHIC, N.D. SOCCI, AND P.G. WOLYNES, Proteins **21**, 167 (1995).

[32] B.W. CHURCH AND D. SHALLOWAY, unpublished results.

GLOBAL OPTIMIZATION FOR MOLECULAR CLUSTERS USING A NEW SMOOTHING APPROACH

C.-S. SHAO* , R.H. BYRD* , E. ESKOW* , AND R.B. SCHNABEL*

Abstract. Strategies involving smoothing of the objective function have been used with some efficacy to help solve difficult global optimization problems arising in molecular chemistry. This paper proposes some new smoothing approaches and examines the utility of smoothing in global optimization. We first propose a new, simple algebraic way of smoothing the Lennard-Jones energy function, which is an important component of the energy in many molecular models. This simple smoothing technique is shown to have close similarities to previously-proposed, spatial averaging smoothing techniques. We then present some experimental studies of the behavior of local and global minimizers under smoothing of the potential energy in Lennard-Jones problems. An examination of minimizer trajectories from these smoothed problems shows significant limitations in the use of smoothing to directly solve global optimization methods. In light of these limitations, a new stochastic-perturbation method that combines smoothing and large-scale global optimization techniques is proposed. A set of experiments with the first phase of this algorithm on Lennard-Jones problems gives very promising results, and offers a clear indication that the use of smoothing in this context is helpful. These smoothing and global optimization techniques are designed to be applicable to a large class of empirical models for proteins.

Key words. Global Optimization, Molecular Chemistry, Lennard-Jones Problems, Smoothing Techniques

1. Introduction. The topic of this research is the development of new smoothing methods for large-scale global optimization problems arising in molecular chemistry applications. Molecular conformation problems give rise to very difficult global optimization problems, and smoothing is a technique that has been used in the chemistry and optimization communities to aid in their solution. Our overall approach is to develop new smoothing techniques that are both effective and inexpensive, and also to consider the integration of smoothing with sophisticated global optimization algorithms.

This paper describes the first stage of this research, the development of new smoothing methods to solve the Lennard-Jones problem. The Lennard-Jones problem an important molecular conformation test problem for two reasons. First, the problem is a very difficult global optimization problem, since it is believed that the number of its minima grows exponentially as $O(e^{N^2})$ [11], and many minima have energy values near the global minimum. Secondly, its objective function, the Lennard-Jones potential energy function, commonly exists within other molecular conformation problems, such as protein folding problems. In part for this reason, the techniques developed in this paper can also be used in smoothing a

* All authors can be reached at Computer Science Department, University of Colorado Boulder, Campus Box 430, Boulder, CO 80309, USA. Research supported by AFOSR Grants No. AFOSR-90-0109 and F49620-94-1-0101, ARO Grants No. DAAL03-91-G-0151 and DAAH04-94-G-0228, and NSF Grant No. CCR-9101795.

wide range of molecular conformation problems including protein folding problems.

The Lennard-Jones problem is to find the minimum energy structure of a cluster of N identical atoms using the Lennard-Jones potential energy. That is, the problem assumes that the potential energy of the cluster is given by the sum of the pairwise interactions between atoms, with these interactions being Van der Waals forces given by the Lennard-Jones 6-12 potential

$$(1.1) \qquad p(r) = \frac{1}{r^{12}} - 2 * \frac{1}{r^6}$$

where r is the distance between two atoms. This potential represents a repulsive-attractive force that is very repulsive at very short distances, most attractive at an intermediate distance, and a very weak attractive force at longer distances. In this formulation, the pairwise equilibrium distance (distance of greatest attraction) is scaled to 1, and the pairwise minimum energy is scaled to -1. If we define the position of the cluster by

$$x = (x_1, x_2, ..., x_N)$$

where x_i is a three dimensional vector denoting the coordinates of the ith atom, then the overall potential energy function is

$$(1.2) \qquad f(x) = \sum_{i=1}^{N} \sum_{j=1}^{i-1} p(d_{ij}) = \sum_{i=1}^{N} \sum_{j=1}^{i-1} \left[\frac{1}{d_{ij}^{12}} - 2 * \frac{1}{d_{ij}^{6}} \right]$$

where d_{ij} is the Euclidean distance between x_i and x_j. We can now denote the problem as

$$(1.3) \qquad \mathbf{LJ} : \min_{x \in D} f(x)$$

where $f(x)$ is function (1.2) and D is some closed region in R^n, $n = 3N$. This problem has been attempted by many different computational approaches, such as [1] [5] [8] [9] [10] [11] [12] [14] [15] [19] [21] [23], that make varying amounts of use of the solution structure of Lennard Jones clusters. A large scale global optimization algorithm that does not utilize the solution structure of the clusters has been developed in past few years by [1], and has successfully solved all Lennard-Jones problems with up to 76 atoms.

In practice, interesting molecular conformation problems contain at least 1,000 to 10,000 atoms, and huge numbers of local minimizers. It is expected that it will generally be too expensive to solve problems of such size by using global optimization algorithms directly. This realization motivates approaches that seek to improve the effectiveness of global optimization algorithms via transformations of the objective (energy) function.

One transformation method is to use a parameterized set of *smoothed* objective functions. The smoothed functions are intended to retain the coarse structure of the original objective function, but have fewer local minimizers. By selecting different smoothing parameters, objective functions with different degrees of smoothness can be derived. The intent is to first solve the global optimization problem on very smooth problems, and then use this solution to gradually solve the problem on less smooth problems and ultimately the original, unsmoothed problem. The connections between the smoothed problems and the original problems are characterized by *trajectories* of minima in the space of the smoothing parameter(s). Along the trajectories, each point is a minimizer of a smoothed problem with some specific smoothing parameter setting. Usually, one end of the trajectory is a minimizer of the original problem. That is, once a minimizer of a smoothed problem is found, its trajectory usually leads to a corresponding minimizer of the original problem. If a problem is *strongly scaled* [16], the trajectory from the global minimizer of a very smoothed problem may lead directly to the global minimizer of the original problem. It is also quite possible, however, that the trajectory from the global minimizer of a very smoothed problem will not lead to the global minimizer of the original problem. Indeed, this paper will show that the latter situation appears to be common in Lennard-Jones problems.

At least three distinct smoothing methods have been proposed and applied to Lennard-Jones problems recently. These are the diffusion equation method [13], the effective energy method [4] [5] [19], and a continuation-based integral transformation scheme [6] [22]. All of the above methods transform the original function into a family of smoothed functions via integration of the original objective function. Such integrations are too expensive to compute at run time. For use in global optimization algorithms, either some approximations must be employed, or look-up tables have to be computed in advance.

In this paper, a new family of smoothing functions is introduced. As opposed to previous methods, the new functions do not use integrations. Instead, the transformation is performed directly in an algebraic form. Another important difference between the new method and previous methods is that our transformation is specially designed to be applied on Lennard-Jones-like functions. We will show, however, that the techniques that are introduced also apply to other important functions, and that they allow our techniques to be applied to many important empirical energy functions, such as the common energy functions for proteins.

In order to utilize the smoothing techniques, a new global optimization method based on smoothing and the algorithm in [1] is introduced in this paper. As in previous approaches using smoothing, the new algorithm contains multiple stages, where each stage uses a less smooth objective function than previous stage does. The original objective function is used in the final stage. Unlike previous approaches, however, the new algorithm incorporates a sophisticated global optimization algorithm at each stage, using the best minimizers from the previous stage as the starting points in the current stage. This enables the method not only to track the trajectories of minimizers from the previous stage, but also to explore newly emerged trajectories and minimizers. In this way, the algorithm is equipped to solve problems that do not have the strong scaling property.

The remainder of the paper describes the new family of smoothing functions, the integration of the global optimization algorithm with smoothing functions, and a variety of experimental results. Section 2 briefly reviews other smoothing methods and how they have been used to solve Lennard-Jones problem. Section 3 introduces the new algebraic smoothing functions and discusses their properties on Lennard-Jones problems, including the results of trajectory-tracking experiments. Section 4 first concisely describes the original global optimization, and then explains how the smoothing functions are integrated into the global optimization algorithm. Section 5 presents some very preliminary results on the use of this algorithm on Lennard-Jones problems. Section 6 gives a brief summary and conclusions.

2. Background on smoothing functions. The basic idea of smoothing is to soften the original function by reducing abrupt function value changes while retaining the coarser structure of the original function. In other words, smoothing dampens high gradient values and fine grain fluctuations in the original function. As a result, nearby minimizers will merge after sufficient smoothing is applied to remove the barriers between them. Therefore, smoothing reduces the total number of minima in the problem.

The Lennard-Jones potential energy function for a pair of atoms has a pole at distance zero, and thus very large derivative values for distances near zero. The pole and large gradient values create huge barriers that separate similarly structured minimizers in the total Lennard-Jones problem. That is a fundamental reason why the Lennard-Jones problem, as well as more complex problems that include the Lennard-Jones potential or similar ones, has so many minima. A Lennard-Jones smoothing technique should remove these barriers in some effective way.

In general, a smoothing technique contains a family of smoothing functions that is parameterized over a smoothing parameter or a set of smoothing parameters. Such a family can be represented as

$$(2.1) \qquad \tilde{f}_s : D \to R, \quad s \in S$$

where D is some closed region in R^n, d is the number of smoothing parameters, and S is some subregion of R^d. By varying the smoothing parameters, one can create a series of functions that gradually smoothes the original function. For the Lennard-Jones problem, a family of smoothed Lennard-Jones problems can be constructed,

$$(2.2) \qquad \mathbf{LJ}_s : \min_{x \in D} \tilde{f}_s(x)$$

where s is some smoothing parameter set and $\tilde{f}_s$ is a smoothed Lennard-Jones potential function. The number of minima is reduced gradually as the objective functions become smoother. It is hoped that often, the smoothed problem will have only one minimizer once the degree of smoothing is sufficiently large.

A general smoothing technique, called spatial averaging, has been studied in various ways in [4] [5] [6] [13] [19] [22]. The fundamental idea of this technique is that the smoothed function value at each point is given by a weighted average of the energy function in a neighborhood of the point using a distribution function centered at this point. The Gaussian distribution function is commonly used. In this case, the smoothing transformation is

$$(2.3) \qquad \tilde{f}_s(x) = \int H(f(x'), s') \cdot e^{\frac{-\|x-x'\|^2}{\lambda^2}} dx'$$

where λ and s' are the smoothing parameters. The parameter λ determines the scale of the Gaussian distribution, while the parameter s' is used with the function H to transform the original function $f(x)$ into a function that has no poles. The transformation $H(f, s)$ is necessary to make the function integrable, and also further dampens the function. In the work of [13] this transformation consists of approximating $f(x)$ by a sum of Gaussian functions, while in the work of [22] the transformation consists of truncating $f(x)$ to some fixed maximum value.

As mentioned in Section 1, varying the smoothing parameters can generate trajectories in configuration space of the smoothing parameters that link minimizers of functions with different amounts of smoothing. Each point on a given trajectory is a minimizer of some smoothed problem. These trajectories can be denoted as

$$(2.4) \qquad traj_i : S \rightarrow D.$$

Since each point on a given trajectory is a minimizer, the points on the trajectory satisfy

$$(2.5) \qquad \nabla_x \tilde{f}_s(traj_i(s)) = 0.$$

In practice, once one has a local minimizer for a given value of the smoothing parameter(s), one can trace its trajectory by slowly changing the smoothing parameter and conducting a series of local minimizations. Each local minimization uses the minimizer from the previous value of the smoothing parameter(s) as the starting point for minimizing the function with the new smoothing parameter(s).

The behavior of these trajectories of smoothed minimizers is the key to understanding the use of smoothing to solve global optimization problems. In the best case, the trajectory containing the global minimizer of highly smoothed problems will also contain the global minimizer of the original problem. In this case, by finding the global minimizer of a very smooth problem (which hopefully is fairly easy to do), and then tracking this trajectory through a careful sequence of local minimizations on less and less smoothed functions, one can locate the global minimize of the original problem. However, this relationship between the global minimizer of very smoothed problems and the global minimizer of the original problem does not always exist. In order to understand the incorporation of smoothing in global optimization, one needs to be aware of the possible pitfalls in trajectory tracking.

The global minimizer of the original problem can fail to lie on the same trajectory as the global minimizer of very smoothed problems for one of at least four reasons. The most important of these is that the order of minimizers may change as the function becomes smoother. For example, as the function is smoothed, a low minimizer that is surrounded by very high barriers may become a higher minimizer than an initially somewhat-higher minimizer that is surrounded by much lower barriers. In this case, the trajectories containing these minimizers will "flip order" at some point. If the trajectory containing the global minimizer of the original problem flips order with another trajectory, then tracking the global minimizer of a very smooth problem back will not lead to the global minimizer of the original problem. The experiments in this paper indicate that this situation appears to be common.

Secondly, any trajectory may terminate beyond a given set of values of the smoothing parameters. This occurs when a minimizer is smoothed away beyond these values of the smoothing parameters. For the trajectory of the global minimizer of the original problem, this can only occur if the trajectory has first "flipped" order as described in the previous paragraph. When such termination occurs, the process of tracking to smoother objectives will generally jump to a different trajectory and continue. If the trajectory from the global minimizer of the original objective terminates in this way, then the global minimizer cannot be found by continuous tracking from a smoother minimizer. A third possibility is the analogous termination of the trajectory from a smoothed minimizer as the parameters make the function less smooth. This less common behavior occurs when smoothing introduces a new minimizer at some point. The experiments in this paper verify that this may occur. If this happens for a trajectory coming from the global minimizer of a highly smoothed problem, no trajectory can lead continuously to the global minimizer of the original problem. Finally, two trajectories may merge as the function becomes more smoothed, if two minimizers merge. This behavior appears to be much less common.

These behaviors are illustrated in Figure 2.1, which plots energy values of several smoothing trajectories that we tracked for a 30 atom Lennard-Jones cluster, using the new smoothing procedure we describe in Section 3. The trajectories show flip order between the original function and the smoothest function, and there is a trajectory from the left and one from the right that terminate so that the tracking drops to a lower trajectory. It is apparent that a method that incorporates smoothing to solve global

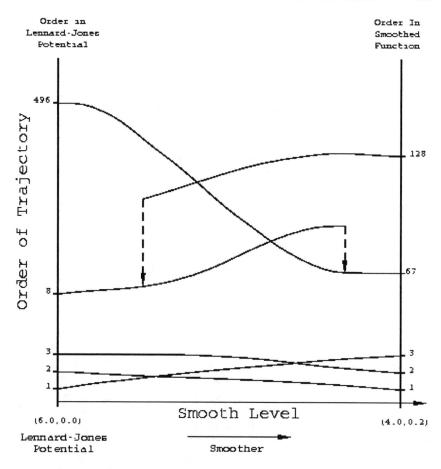

FIG. 2.1. *Trajectories of selected minimizers for 30 Atom problem.*

optimization problems will need to deal with all these behaviors. The numbers in parentheses, (6.0,0.0) and (4.0,0.2), are the smoothing parameters of our new smoothing function family which is discussed in next section.

Related behaviors to those just described have been reported in the work of [16] on state trajectories. State trajectories are intended to capture the transitions between *valleys* of minimizers in the smoothing hyperspace. Three types of patterns are described in [16], *strong scaling, weak scaling,* and *poor scaling*. These correspond to the abilities of trajectories of smoothed problems in the formulation of [16] to lead to the original global minimizer, with strong scaling corresponding to the "best" case mentioned above.

Several global optimization algorithms and smoothing techniques have been coupled and applied on Lennard-Jones problem with partial success. The preceding discussion of trajectory tracking is helpful in describing these methods. The diffusion equation method [13] treats the original Lennard-Jones objective function as the initial condition of a temperature field whose governing equation is a diffusion equation. There is a time scale that controls the smoothness of the field. The field becomes homogeneous as the time scale approaches infinity. The method first solves the problem at some large time scale where the problem either has only one minimizer or very few. Then, it tracks the minimizer(s) back to time scale zero by a sequence of local optimizations. That is, a trajectory tracking procedure is applied on each of a small number of minima of the very smoothed problem. The effective energy method [4] [5] [19] tries to simulate the physical annealing process with an effective energy function. The effective energy function is a function of a temperature variable and a spatial scale. When the spatial scale reduces to zero and the temperature drops to some low temperature where the stable state is reached, the effective energy function returns to the Lennard-Jones potential function. This method uses the traditional simulated annealing algorithm at each smoothing stage. It does not track any trajectories. The continuation-based integral transformation scheme [6] [22] directly applies a spatial averaging equation like (2.3) to the objective function. Both methods that apply a global algorithm such as simulated annealing at each stage, and methods that track trajectories, have been considered in conjunction with this approach.

In the next section, we will discuss a new family of smoothing functions for Lennard-Jones problems. As opposed to spatial averaging techniques, this family of smoothing functions does not involve integration which first requires modification of the Lennard-Jones potential. Instead, this family of functions removes the poles from the Lennard-Jones function in a continuous, algebraic manner. In section 4, we will discuss the combination of this (or other) smoothing approach with more sophisticated global stages than are used in the work mentioned above.

3. A new family of smoothing functions. In this section we introduce a new, algebraic method of smoothing the Lennard-Jones potential energy function. Next we present the results of experiments that show how this smoothing method compares to the integration-based smoothing approach of [6] [22]. Finally we present the results of experiments that illustrate the effect of the new smoothing approach on the energy surface of Lennard-Jones clusters. While the smoothing function and experiments presented in this section are limited to Lennard-Jones clusters, we will comment later in this paper about how these techniques will allow us to smooth more complex energy functions including empirical potentials for proteins.

The new family of smoothing functions we propose for the Lennard-Jones potential is

$$(3.1) \qquad \tilde{p}(r, P, \gamma) = \left(\frac{1 + \gamma}{r^P + \gamma} \right)^2 - 2 * \left(\frac{1 + \gamma}{r^P + \gamma} \right)$$

where r is the distance between atoms, and γ and P are the two smoothing parameters. This equation is equal to the original Lennard-Jones potential (1.1) when $\gamma = 0$ and $P = 6$. For any values of $\gamma >= 0$ and $P > 0$, it attains its minimum value of -1 at $r = 1$, as does the Lennard-Jones potential. Furthermore, equation (3.1) only contains simple algebraic computations, and is nearly as inexpensive to evaluate as the original Lennard-Jones potential (1.1).

The two smoothing parameters in equation (3.1) serve two different roles. As P becomes smaller than six, the function becomes smoother and the basin around its minimizer becomes broader. Fig. 3.1 illustrates this, showing the potential between two atoms for different values of P while γ is fixed at zero. The values of the function near the equilibrium distance of one clearly are reduced as P decreases. However, the function value still goes to infinity as the distance approaches zero.

The smoothing parameter γ is used to remove the pole from the Lennard-Jones potential. Fig. 3.2 shows the smoothed function for various values of γ, with $P = 6$ in all cases. This figure illustrates that the y-intercept of the smoothing function, $\tilde{p}(0, P, \gamma)$, decreases as γ increases. In addition, increasing γ reduces the values of the function for $r < 1$, but has minimal effect for $r > 1$. The relation between γ and $\tilde{p}(0, P, \gamma)$ can be easily obtained from equation (3.1) and is given by

$$(3.2) \qquad\qquad \tilde{p}(0, P, \gamma) = \frac{1}{\gamma^2} - 1.$$

It should be noted that the relation (3.2) is not a function of P and that $\tilde{p}(0, P, \gamma)$ approaches -1 as γ approaches infinity. In addition, when $\gamma > 0$ and $P > 1$, equation (3.1) has a zero derivative at $r = 0$. Fig. 3.3 clearly illustrates the independence of the y-intercept upon the parameter P, as well as the zero derivative at $r = 0$.

While preparing this paper for publication, we discovered a recent paper [17] that proposes a smoothing function quite similar to (3.1). They propose a smoothing function where r in (1.1) is replaced with $(r+\alpha)/(1+\alpha)$ for some positive constant α. Note that our smoothing function (3.1) may be viewed as replacing r^6 in (1.1) by $(r^P + \gamma)/(1 + \gamma)$.

It would be interesting to compare our new smoothing functions with the spatial-averaging smoothing techniques applied to the Lennard-Jones function. However, there is one fundamental difference between our smoothing functions and spatial-averaging smoothing functions. It is that the minima of equation (3.1) are fixed at $r = 1$ with function value -1, while the minima of the spatial-averaging smoothing functions drift outward with function value higher than -1 as the potential function becomes smoother. Fig. 3.4 shows the spatial-averaging smoothing functions in [6] [22] for different smoothing parameter values. These functions illustrate this effect. Note that to generate these functions, the Lennard-Jones function first must be truncated for values of r close to zero.

Therefore, in order to obtain a quantitative comparison of the new

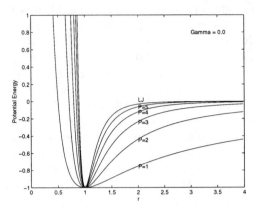

FIG. 3.1. *Potential Energy Curves Between Two Atoms on Different Settings of P at* γ = 0.0

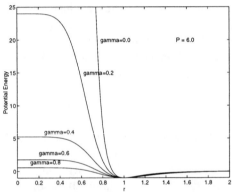

FIG. 3.2. *Potential Energy Curves Between Two Atoms on Different Settings of* γ *at P* = 6.0

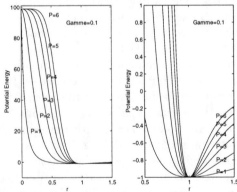

FIG. 3.3. *Potential Energy Curves Between Two Atoms on Different Settings of P at* γ = 0.1

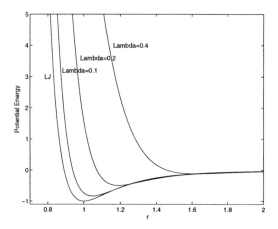

FIG. 3.4. *Potential Energy Curves Between Two Atoms for Spatial-Averaging Smooth Functions in [6][22].*

smoothing function with spatial-averaging smoothing functions, two scaling parameters are temporarily added into equation (3.1). This gives the equation

$$(3.3) \qquad \tilde{p}(r, P, \gamma, a, b) = b * \{ (\frac{1 + \gamma}{(a \cdot r)^P + \gamma})^2 - 2 * (\frac{1 + \gamma}{(a \cdot r)^P + \gamma}) \}$$

where a and b are the new scaling parameters. These two parameters only change the scale of the potential function (3.1), and do not change its smoothing properties or the relative locations of minimizers. In terms of the underlying molecular configurations, the parameter a expands the configuration, while the parameter b multiplies the energy value by a constant without changing the shape of the configuration.

With the new scaling parameters, we can compare the two families of smoothing functions by curve fitting them directly. The least squares method was used to find the parameter settings for the function (3.3) that allow it to match most closely the spatial-averaging smoothed Lennard-Jones functions of [6] [22] for different values of their smoothing parameter λ. The fitting is performed over the "interesting" ranges for the distance parameter r, that is omitting values of r so small that they never occur in realistic clusters, or so large that the energy value is negligible, and over a grid of points with spacing of .01. Since these curves' minima tend to shift to the right as *lambda* increases, our fitting ranges shift also. The ranges are shown in Table 3.1, which lists the corresponding parameters and residuals. The residual shown is the square root of the average squared residual over the grid points. The low residuals show that our family of functions can match the spatial-averaging functions extremely closely over these fitting ranges. Fig. 3.5 demonstrates the similarity between the two

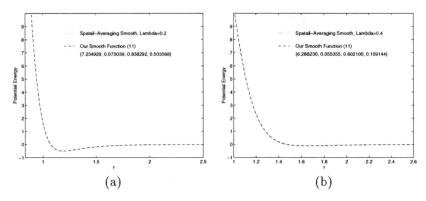

FIG. 3.5. *Comparison of the spatial-averaging smoothing functions in [6][22] with our smoothing functions (3.3): (a) curve fitting the spatial-averaging smoothing function at* $\lambda = 0.2$ *with our smoothing function at* $(P = 7.234929, \gamma = 0.075039, a = 0.838292, b = 0.503568)$ *over* $[0.9, 2.5]$; *(b) curve fitting the spatial-averaging smoothing function at* $\lambda = 0.4$ *with our smoothing function at* $(P = 6.286230, \gamma = 0.055355, a = 0.602100, b = 0.109144)$ *over* $[1.0, 2.6]$.

families of smoothing functions. At this resolution, one cannot perceive any difference between the shapes of the two functions.

TABLE 3.1

Results of least squares fitting function (3.3) with function curves from [6][22] for different values of λ.

λ	Range	P, γ, a, b	RMS Residual
0.1	$[0.83, 2.5]$	6.348, 0.008, 0.953, 0.844	0.00674
0.2	$[0.90, 2.5]$	7.235, 0.075, 0.838, 0.504	0.00773
0.3	$[0.94, 2.6]$	6.923, 0.071, 0.712, 0.235	0.00984
0.4	$[1.00, 2.6]$	6.286, 0.055, 0.602, 0.109	0.01647

The ability of our new class of smoothing functions to so closely match the spatial-averaging smoothed Lennard-Jones functions is somewhat surprising, since the techniques used to derive the two sets of smoothing functions are so different. These results lead us to expect that our new smoothing functions should be at least as effective as the spatial averaging smoothing function in global optimization algorithms for Lennard-Jones problems. In addition, our new smoothing function family is easier to evaluate and implement, and the family offers a broader spectrum of smoothing functions with its two smoothing parameters. Given the similarity between the curves in Fig. 3.5, one would expect that the two families should have very similar general properties, such as order flips, at equivalent smoothing levels. We do not make further comparisons between the two families as we explore smoother levels for the two-parameter family, because it appears difficult or impossible to find equivalent smoothing levels in the one-parameter spatial-averaging family.

It is worth noting that in Table 3.1, the value of P first rises above 6, up to 7.2, while the value of γ is generally small. This indicates that the spatial-averaging technique may make the problem more difficult than the original problem for small values of the smoothing parameter. This effect appears to come from the pole in the Lennard-Jones function, which is cut off at a fairly high level in the approach of [6] [22]. When the smoothing parameter is small, these high values have a large effect for smaller values of r, and drive the smoothing function up sharply in this range. This effect is lessened as the smoothing parameter becomes larger so that the integration gives more weight to more distant function values. Probably, this effect also could be lessened by using different cut-off values for the Lennard-Jones pole prior to the integration.

Now, we can define the new smoothed Lennard-Jones objective function by

$$(3.4) \qquad \tilde{f}_{<P,\gamma>}(x) = \sum_{i=1}^{N} \sum_{j=1}^{i-1} \tilde{p}_{<P,\gamma>}(d_{ij})$$

and the smoothed Lennard-Jones problem by

$$(3.5) \qquad \mathbf{LJ}_{<P,\gamma>} : \min_{x \in D} \tilde{f}_{<P,\gamma>}(x).$$

In order to examine the effect of the new smoothing functions in detail, we first tried to solve the nine-atom Lennard-Jones problem comprehensively, using the global optimization method of [1] that is reviewed in the next section. First, we were surprised to find twenty distinct minimizers, instead of the eighteen minimizers that are reported in [10] and have often been cited in the literature. This demonstrates the difficulty of solving the Lennard-Jones problem comprehensively. Secondly, the trajectories of all twenty minimizers were tracked carefully through different values of the smoothing parameters P and γ. The tracking procedure is a series of local minimizations with objective function (3.4) through a sequence of smoothing parameter sets, $\{(P_0, \gamma_0), (P_1, \gamma_1), (P_2, \gamma_2), ...\}$; where $P_0 = 6$, $\gamma_0 = 0$ and (P_{i+1}, γ_{i+1}) is more smooth than (P_i, γ_i). The local minimizations were performed using the BFGS method of the UNCMIN package [18]. Great care was taken to assure that the step sizes in the smoothing parameter space are small enough that the tracking procedure is stable. If the step size is too large, the tracking process may "jump" off of the current trajectory to another trajectory due to the large change in the potential energy landscape. A step size of 10^{-3} was used for both P and γ in these experiments. For different values of the smoothing parameters, the number of distinct minimizers that exist is recorded in Table 3.2. This number represents the number of the original trajectories that still exist at this smoothing level.

TABLE 3.2

Number of trajectories with different values of smoothing parameters. A number with bold type denotes that the trajectory of original global minima is no longer global under the smooth setting.

γ	$f(0)$	P = 6.0	P = 5.0	P = 4.0	P = 3.0	P = 2.0
0.0	inf	20	16	11	6	4
0.1	99.0	16	15	9	6	1
0.2	24.0	16	12	7	5	1
0.3	10.1	16	10	6	4	1
0.4	5.25	14	8	5	1	1
0.5	3.0	11	6	5	1	1
0.6	1.78	10	5	2	1	1
0.7	1.04	8	5	1	1	1
0.8	0.56	6	3	1	1	1
0.9	0.23	5	1	1	1	1

The steady reduction in the number of minimizers shown in Table 3.2 clearly shows that the new smoothing function (3.4) effectively smoothes the Lennard-Jones problem. The experiment also demonstrates the transitions of trajectories through smoothing parameter space. For example, four trajectories have either terminated or merged with other trajectories between $(P = 6, \gamma = 0.0)$ and $(P = 5, \gamma = 0.0)$, or between $(P = 6, \gamma = 0.0)$ and $(P = 6, \gamma = 0.1)$. Once a certain level of smoothing is reached or exceeded, only one minimizer is left, and all remaining trajectories have terminated.

Table 3.2 further indicates how the smoothing technique could be used for solving global optimization problems. For example, for the nine-atom Lennard-Jones problem, one could first solve the smoothed problem at $P = 5$ and $\gamma = 0.5$, where there are only six instead of twenty minimizers. Then, one could either find the global minimizer of $\mathbf{LJ}_{<5,0.5>}$ and perform the reverse tracking procedure on it, or find all minimizers in $\mathbf{LJ}_{<5,0.5>}$ and track their trajectories backward. One problem with this approach is that the number of minima of the Lennard-Jones problem, and other molecular configuration problems, grows exponentially. For example, for the 15 atom Lennard-Jones problem, the number of minima is estimated to be more than 10,000. Suppose that the smoothed potential function for $(P = 5, \gamma = 0.5)$ reduces the number of minima at the same rate as for the 15 atom problem. Then there would still be 3,000 minimizers in $\mathbf{LJ}_{<5,0.5>}$ for 15 atoms. This illustrates the improbability of solving large global optimization problems that have exponential numbers of minimizers by finding all the minimizers, even of smoothed functions. This observation serves as part of the motivation for the combination of global optimization techniques with smoothing techniques that is discussed in the next section.

Most importantly, however, Table 3.2 clearly illustrates one of the biggest limitations of the smoothing approach, *order flips*. This means that the orders of trajectories can change through smoothing parameter space. That is, at a given smoothing level, a given trajectory can have a higher smoothed minimizer than a second trajectory, even though the initial unsmoothed minimizer that started the first trajectory had a lower function value than the unsmoothed minimizer that started the second trajectory. In the worst case, the global minimizer's trajectory can drop its order dramatically as it is smoothed. If this happens, this means the global minimum can not be found by simply tracking the lowest minimizers of very smoothed problems backwards. In table 3.2, the smoothing levels where the global minimizer's trajectory has dropped its order have been denoted in boldface. It is seen that for this problem, this phenomenon *always* occurs by the time there are five or fewer trajectories remaining.

A more detailed analysis of the results of the nine-atom trajectory tracking experiment reveals that the trajectory of the unsmoothed global minimizer first drops in order to second place once a certain level of smoothing is reached. Then, at some greater value of smoothing, it terminates and local minimizations from it jump into the current smoothed global minimizer's trajectory. This trajectory is the one that originates from the second lowest minimizer of the original problem. A reverse tracking procedure has been performed on the only minimizer remaining at $(P = 2, \gamma = 0.9)$. There is no surprise that the reverse tracking procedure leads to the second lowest minimizer of the original problem, instead the global minimizer. This simple problem illustrates that if the global minimizer flips its order, the global minimum can not be found by only reverse tracking the best smoothed minimizer, and if the global minimizer's trajectory terminates, the global minimum can not be found by reverse tracking any trajectories from very smoothed problems. In spite of these limitations, we feel that smoothing has much to offer in difficult large global optimization problems, and illustrate this in the next two sections of this paper.

To examine the order flip phenomenon further, we next tested the Lennard-Jones problem with thirteen atoms. We started by trying to find all the minimizers of the unsmoothed problem, again using the global optimization algorithm of [1] and found 1,044 distinct minimizers. Then, we tracked all 1,044 minimizers' trajectories from the original function to smoothing level $(P = 4.0, \gamma = 0.2)$. The result is shown in Fig. 3.6a and Fig. 3.6b. The horizontal axis represents the trajectory's order in the Lennard-Jones potential, and the vertical axis represents the order of the minimizer on this trajectory for smoothing function $\tilde{f}_{<4.0,0.2>}$. If a trajectory terminates before smoothing level $(P = 4.0, \gamma = 0.2)$ (or merges with another trajectory), its will have the same smoothing order as the trajectory which it jumps down to (or merges with). Fig. 3.6a shows that only 26 trajectories are left by smoothing level $(4.0, 0.2)$. Most of the remaining trajectories have terminated before this smoothing level and jumped down into the global minimizer's trajectory. Fig. 3.6b is a blow-up of Fig. 3.6a for the best 50 original minimizers. In this case, there are not many order flips. The first flip is between the seventh and nineteenth minima in the original Lennard-Jones order, which have smoothing orders six and five respectively. These results shows that smoothing level (4.0, 0.2) is already very smooth for the 13-atom problem, and that, by reverse tracking, we can determine the unsmoothed global minimizer from the smoothed global minimizer. In addition, the trajectory ordering diagram indicates that for this problem, apparently most of minimizers are in a broad basin whose lowest point is the global minimum of the original problem. In addition, the small number of order flips indicates that the energy landscape of the potential function does not change dramatically as the function becomes smoother. However, the 13-atom problem is one of the easier Lennard-Jones problems because the solution has a full outer shell and is completely symmetric. This may well lead to a nicer correspondence between the unsmoothed and smoothed problems.

Next, we examined the order flip phenomenon for the Lennard-Jones problem with 30 atoms. In addition to being substantially larger than the previous cases, this problem does not have a symmetric, full outer-shell solution like the 13 atom problem. We began by locating 1,043 of the lowest minimizers for this problem, including the global minimizer, using our global optimization algorithm discussed in the next section. Unlike the 9-atom and 13-atom problems, it is not feasible to find all the minimizers for the 30-atom problems. Indeed, according to [10], we have found only a tiny portion of all the minimizers. All 1,043 minimizers were tracked to the smoothing level (4.0,0.2) by the same tracking procedure used in the previous experiments. The order flip graphs are shown in Fig. 3.7a and 3.7b.

The order flips in the 30-atom problem are much worse than the flips in the 13-atom problem. This experiment clearly demonstrates that order flips among trajectories is a significant issue using smoothing. Fig. 3.7b shows that the trajectory of the original global minimizer has dropped down to third place by smoothing level (4.0,0.2), More dramatically, the minimizers in fourth and sixth to eighth positions in the original function are all below twentieth place by this smoothing level. Similarly, the trajectories which are in seventh to tenth positions for smoothing level (4.0,0.2) were all below twentieth place for the original function. This means that if we run a global optimization algorithm at this smoothing level and then simply track the trajectories of the best smoothed minimizers back to the original function, we will miss some of the best Lennard-Jones minimizers and will find several poor Lennard-Jones minimizers. Furthermore, since the original global minimizer has already dropped in order, it is possible that its trajectory will terminate at some further smoothing level and that if we start reverse tracking from that level we could miss the original global minimizer entirely.

In Fig. 3.7, almost half of the minimizers' trajectories have terminated

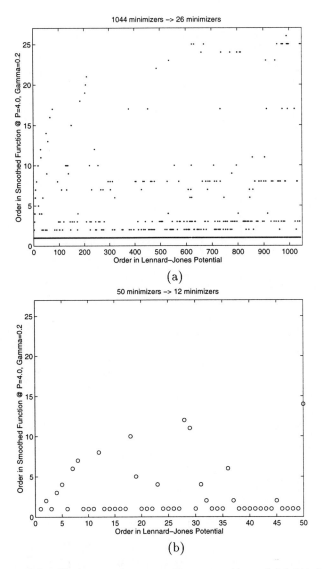

FIG. 3.6.   *Order flipping experiment on 13-atom problem: (a) 1,044 Lennard-Jones minimizers are tracked to smoothing level $P = 4, \gamma = 0.2$. The horizontal axis represents the trajectory's order in Lennard-Jones potential, and the vertical axis represents the trajectory's order in smoothing function, $\tilde{f}_{<4.0,0.2>}$. (b) The first twenty minimizers of Fig. 3.6a.*

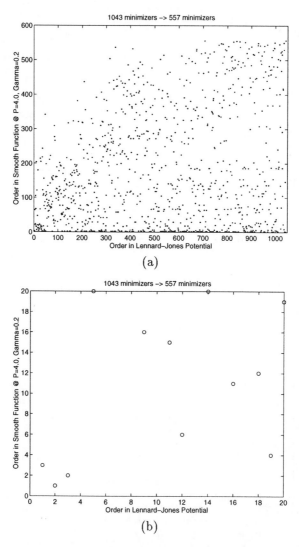

FIG. 3.7. *Order flipping experiment on 30-atom problem: (a) 1,043 Lennard-Jones minimizers are tracked to smoothing level $P = 4, \gamma = 0.2$. The horizontal axis represents the trajectory's order in Lennard-Jones potential, and the vertical axis represents the trajectory's order in smoothing function, $\tilde{f}_{<4.0,0.2>}$. (b) The first twenty minimizers of Fig. 3.7a.*

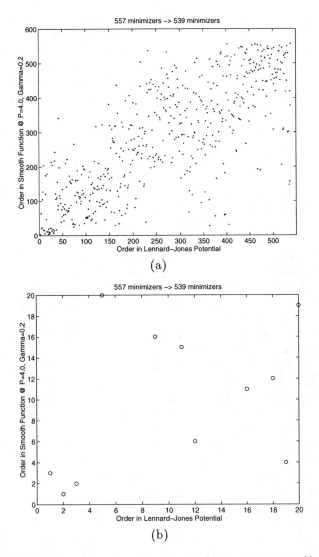

FIG. 3.8. *Order flipping experiment with reverse tracking on 30-atom problem: (a) 557 minimizers of smoothing level $P = 4, \gamma = 0.2$ are tracked back to Lennard-Jones potential. The horizontal axis represents the trajectory's order in Lennard-Jones potential, and the vertical axis represents the trajectory's order in smoothing function, $\tilde{f}_{<4.0, 0.2>}$. (b) The first twenty minimizers of Fig. 3.8a.*

and collapsed into other minimizers' trajectories. From the data in these kinds of figures, we can not tell which original trajectories have terminated and which have led directly to the minimizers of the smoothed problem, since there is no guarantee that the trajectory with the lower original minimizer is the one that exists longest. Thus, we cannot tell from these figures which original minimizers will be found by reverse tracking from the smoothed minimizers. Since this is a crucial piece of information in assessing how smoothing will help solve global optimization problems, we have conducted a second experiment for the 30-atom problem using reverse tracking.

To perform the reverse tracking experiment, we collected all 557 distinct smoothed minimizers remaining in the 30-atom problem at smoothing level (4.0,0.2), and applied the same tracking procedure described above in reverse to reach the original Lennard-Jones potential. The result is shown in Fig. 3.8. By removing the minimizers whose trajectories terminate during forward tracking, Fig. 3.8a clearly confirms the phenomenon of order flipping in the smoothed 30-atom Lennard-Jones problem. That is, the reverse tracking experiment also produces a multitude of large order flips. In addition, Fig. 3.8b shows that the same pattern of flips is observed for the lowest minimizers in the reverse tracking experiment as in the forward tracking experiment. Taken together, these experiments indicate that order flips are likely to be an important issue to consider in using smoothing in global optimization algorithms. Another interesting observation from the reverse tracking experiment is that some trajectories terminate during the reverse tracking process. In this experiment, 18 of the 557 trajectories terminate before the original problem is reached. This termination as the problem becomes less smooth confirms one of the types of behaviors of smoothing trajectories that was mentioned in Section 2, and implies the emergence of new trajectories as the problem becomes smoother. After such a trajectory terminates during reverse tracking, the tracking procedure will falls to another, lower trajectory. This behavior is far less common than trajectories terminating during forward tracking, since it only occurs when a new minimizer *emerges* as the function becomes smoother, while smoothing tends to reduce the number of minimizers overall.

Finally, we conducted the same experiments for the 34-atom Lennard-Jones problem as for the 30-atom problem. The 34 atom problem appears to be one of the most difficult Lennard-Jones problems among the first 75; the global minimizer, which has an energy value of -150.045, appears to be in a much narrower and more isolated basin of attraction than the second lowest minimizer with energy value -149.997. For this problem we began by locating 1288 of the lowest minimizers and tracking them to smoothing level (4.0, 0.2), resulting in 494 distinct minimizers at this smoothing level. Then we tracked these 494 minimizers in reverse back to the original Lennard-Jones function, resulting in 461 minimizers. The correspondence between the unsmoothed and smoothed minimizers when tracked in these

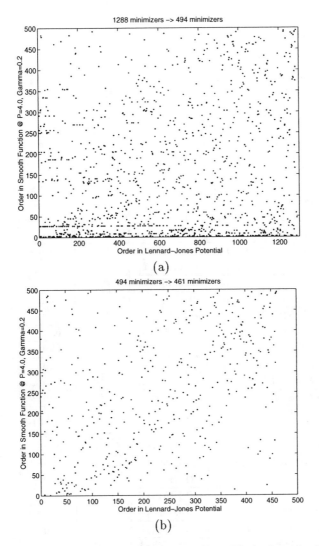

(a)

(b)

FIG. 3.9. *Order flipping experiment on 34-atom problem: The horizontal axis represents the trajectory's order in Lennard-Jones potential, and the vertical axis represents the trajectory's order in smoothing function,* $\tilde{f}_{<4.0,0.2>}$. *(a) 1288 Lennard-Jones minimizers are tracked to smoothing level* $(P = 4, \gamma = 0.2)$. *(b) 494 minimizers of smoothing level* $(P = 4, \gamma = 0.2)$ *are tracked back to corresponding Lennard-Jones potential minimizers.*

TABLE 3.3

Results of tracking best 20 Lennard-Jones minimizers to minimizers of smoothing level $(P = 4.0, \gamma = 0.2)$, *and results of tracking smooth minimizers that lead back to best 20 Lennard-Jones minimizers.*

Trajectory tracking status: "Jump" - trajectory terminates and falls into other trajectory. "Cont" - trajectory is tracked to smoothing level (4.0,0.2) without termination. "None" - did not find trajectory leading to specified Lennard-Jones minimizers.

Best 20 of 1288 Original Lennard-Jones Minimizers		Best 20 LJ Min. → Smooth Min. @ (4.0,0.2)			Smooth Min. @ (4.0,0.2) → Best 20 LJ Min.	
		Trajectory Tracking Status	Final Smoothed Minimizer	Order in Smoothed Potential	Trajectory Tracking Status	Order in Smoothed Minimizers
1	-150.044528	Jump	-215.667165	135	Jump	241
2	-149.997007	Jump	-216.732431	27	Jump	381
3	-149.921622	Cont	-215.300955	205	Cont	205
4	-149.886921	Cont	-214.933377	292	Cont	292
5	-149.795387	Cont	-215.088910	257	Cont	257
6	-149.780940	Jump	-218.359634	3	None	
7	-149.672121	Cont	-218.390471	2	Cont	2
8	-149.494653	Jump	-216.732431	27	None	
9	-149.482330	Cont	-215.103792	250	Cont	250
10	-149.470798	Jump	-218.359634	3	None	
11	-149.434941	Cont	-214.873379	305	Cont	305
12	-149.223885	Cont	-215.060548	267	Cont	267
13	-149.162201	Jump	-214.844140	307	None	
14	-149.095462	Jump	-215.300955	205	None	
15	-149.093101	Cont	-215.427124	180	Cont	180
16	-149.091680	Jump	-215.101311	252	None	
17	-149.082624	Cont	-213.391740	483	Cont	483
18	-149.058111	Jump	-216.732431	27	None	
19	-149.054405	Cont	-213.646593	479	Cont	479
20	-149.010289	Cont	-219.480360	1	Cont	1

two directions is shown in Figures 3.9a and 3.9b. These figures show that the order flip phenomenon is even more pronounced for this problem than for the 30-atom problem, and is really quite extreme. Table 3.3 summarizes the results of these tracking experiments for the best minimizers of the unsmoothed problem. For each of the twenty lowest Lennard-Jones minimizers, after tracking the smoother function, the objective value of the smoothed minimizer reached is shown in column 4 and its order (among the 494 found)in column 5. In column 3, the designation *Jump* indicates the trajectory of the unsmoothed Lennard-Jones minimizer terminated, and the tracking procedure jumped to another trajectory, while *Cont* indicates a single continuous trajectory was followed. It is seen that almost all of the 20 lowest Lennard-Jones minimizers, including the global, tracked to relatively poor smoothed minimizers. In the reverse tracking from the 494 smoothed minimizers, we determined which of these minimizers led to one of the best twenty Lennard-Jones minimizers. For each of these best 20, column 7 gives the order of the lowest smoothed minimizer from which reverse tracking led to it. If the trajectory from the smoothed minimizer terminated during reverse tracking, and the tracking procedure jumped to another trajectory, column 6 says *Jump*; if the trajectory went all the way, the entry is *Cont*. The entry *None* means that no smoothed minimizer was found that would track to that unsmoothed minimizer. It is seen from column 7 that only 2 of the 20 lowest smoothed minimizers lead to any of the best 20 unsmoothed minimizers via reverse tracking, and that the 241st lowest smoothed minimizer is the best one that leads back to the global minimizer, though with a trajectory jump. Thus, reverse tracking 20 lowest smoothed minimizers we found would not find the global minimizer, and would find only two of the best 20.

These experiments show that following trajectories is not completely reliable, and it is necessary to deal with various potential pitfalls of the smoothing process when using it in a global optimization algorithm. In next section, a new global optimization algorithm that utilizes smoothing is presented. The new algorithm integrates smoothing techniques into the general global optimization algorithm of [1]. The algorithm is designed so that it can utilize the ability of smoothing functions to find minimizers on good trajectories easily, while also utilizing the ability of the global optimization algorithm to find better minimizers from good minimizers at any stage of the algorithm. This characteristic enables the algorithm to overcome the pitfalls of the smoothing process that this section has illustrated.

4. Integration of smoothing technique and global optimization algorithm. In this section, a new global optimization algorithm is discussed. The new algorithm is designed to utilize smoothing functions for solving the Lennard-Jones problem. It is based on the algorithm proposed in [1], which has been successful in solving Lennard-Jones problems with up to 76 atoms.

A global optimization method that utilizes smoothing functions has to consider the possible problematic behaviors of trajectories that were illustrated in the previous section. Therefore, a robust method must be more than just optimization of a smoothed function followed by a trajectory tracking procedure. For one thing, it needs to successfully overcome trajectory terminations and order flips. Our algorithm deals with these issues by including a global optimization procedure at several stages of the algorithm. This procedure allows the algorithm to utilize good smoothed minimizers to locate related but even better minimizers that are not necessarily on any trajectory from the previous level of smoothing. In essence, each level of smoothing is used to provide excellent starting points for global optimization of the next, less smoothed function.

This section first briefly describes the algorithm proposed in [1], which is the basis for the global optimization techniques that we use along with smoothing. Then we discuss the integration of the global optimization algorithm and the smoothing techniques. Both algorithms are described in the context of the Lennard-Jones problems. However our global optimization algorithm has been used on other molecular conformation problems, including water clusters [2] and proteins [7], and the new global optimization approach with smoothing also generalizes to a broad class of problems.

The global optimization algorithm in [1] consists of two phases. The first, sample generation phase uses random sampling and local minimizations in the full-dimensional variable space to generate a set of low local minimizers. The second phase generates improved local minimizers from the current set of local minimizers. This phase constitutes the major portion of the computational effort of the method. Each phase contains a procedure that works on small-dimensional subproblems within the full dimensional problem, taking advantage of the partial separability of the objective function to make this efficient. (Partial separability means that the objective function is the sum of functions, each one of which involves only a small number of the variables, in this case two atoms. Because of the partial separability of the Lennard-Jones function, re-evaluating it when only one atom changes costs $2/N$ of what a full function evaluation costs.)

The first phase of the algorithm starts by randomly generating a moderate sized sample of configurations in a sampling domain. For each atom, this domain consists of a three-dimensional rectangle, in which the atom is generated independently with a uniform distribution. Then the first new small-subproblem procedure, the generation of improved sample points through one-atom sampling, is applied as follows. First, all but a small fraction of the sample points with the lowest function values are discarded. Then each remaining sample point whose function value is above some threshold value is perturbed in the following manner. The contribution of each atom in the sample point to the total function value of that point is calculated, and the atom that contributes the most to the overall function value is selected. This atom is then randomly resampled some number of times, and for each new sample value of this atom, the function value of this molecular configuration with the remaining atoms unchanged is calculated. The coordinates for the resampled atom that result in the lowest function value are retained as the new values of these variables within this sample point. If the function value of the modified sample point is still above the threshold value, the process is repeated, i.e. a (new) single atom is selected, resampled, and moved, until the threshold value has been reached. The threshold value may be varied slightly for each sample point, to ensure variation within the sample. From these perturbed sample points, start points for local minimizations are selected and a local minimization is performed from each of these start points (using a quasi-Newton method). The k lowest local minimizers found from these start points are passed on to Phase 2. The process described above has been found to lead to low sample points and low local minimizers far more efficiently than just sampling in the full dimensional space.

In the second phase, a second new small-dimensional subproblem procedure is used to improve the function values of already low local minimizers through one-atom *global* optimizations. A heuristic (described below) is used to select a configuration for improvement from among the low local minimizers, as well as an atom within that configuration to be moved. Then, rather than just sampling on the selected atom as in the sample point improvement procedure in Phase 1, a complete global optimization algorithm is applied to find the best new positions for the selected atom within the selected configuration, with the remainder of the configuration temporarily fixed. The global optimization method used is the stochastic method of [3] [20], which is a very effective global optimization method for problems with very small numbers of variables. In this case, there are three variables, the coordinates of the selected atom. Generally, these best new positions found for the selected atom by the small global optimization algorithm lead to configurations that are close to, but not exactly, local minimizers for the entire molecule. Therefore a full-dimensional local minimization algorithm is then applied to "polish" the few best of these new configurations. (We chose the few best rather than just the best because we have found that sometimes the best polished solution does not come from the best unpolished solution.) These new full-dimensional local minimizers are then merged with those found previously, and the process just described – select a minimizer, select an atom from it, one-atom global optimization, full-dimensional local minimization polishes – is repeated a fixed number of times. This phase is able to identify significantly improved local minimizers and leads to the success of the method.

The heuristic used to determine which configuration and atom are selected at each iteration of the second phase is the following. We consider an initial configuration and any configurations generated from it (via global

Algorithm 1 – Framework of the Large-Scale Global Optimization Algorithm for Lennard-Jones Problems

1. **Sample Generation Phase**

   a. **Sampling in Full Domain :** Randomly generate the co-ordinates of the sample points in the sampling domain, and evaluate $f(x)$ at each new sample point. Discard all sample points whose function value is above a global "cutoff level".

   b. **One-atom Sampling Improvement :** For each remaining sample point : While the energy of the sample point is above the threshold value, Repeat:
   - Select the atom that contributes most to the energy function value.
   - Randomly sample on the location of the selected atom.
   - Replace this atom in the sample point with the new sample coordinates that give the lowest energy value.

   c. **Start Point Selection :** Select a subset of the improved sample points from step 1b to be start points for local minimizations.

   d. **Full-Dimensional Local Minimizations :** Perform a local minimization from each start point selected in step 1c. Collect some number of the best of these minimizers for improvement in Phase 2.

2. **Local Minimizer Improvement Phase:** For some number of iterations:

   a. **Select a Configuration :** From the list of full-dimensional local minimizers, select the local minimizer and atom to be optimized using the heuristic described in Sub-algorithm 2.

   b. **Expansion :** Transform the configuration by multiplying the position of each atom relative to the center of mass of the configuration by a constant factor of between 1.0 and 1.75.

   c. **One Atom Global Optimization :** Apply a global optimization algorithm to the expanded configuration with only the atom chosen in sub-algorithm 2 as a variable.

   d. **Full-Dimensional Local Minimization :** Apply a local minimization procedure, using all the atoms as variables, to the lowest configurations that resulted from the one-atom global optimization.

   e. **Merge the New Local Minimizers :** Merge the lowest new configurations into the existing list of local minimizers.

Sub-algorithm 2 – Heuristic for selecting a configuration – step 2a of Algorithm 1

1. **Balancing Phase:**   For some number of iterations:
   a. Choose the set containing an initial minimizer plus all if its descendants which has been used the least number of times so far.
   b. Choose the configuration from this set with the lowest energy function value, and which has previously been worked on less than two times. If this set contains no such configurations, go back to step a.
   c. If this configuration has never been worked on then choose the atom with the largest energy contribution, otherwise choose the atom with the second largest energy contribution.

2. **Non-balancing Phase:**   For some number of iterations:
   a. Choose the configuration with the lowest energy function value which has previously been worked on less than two times.
   b. If this configuration has never been worked on then choose the atom with the largest energy contribution, otherwise choose the atom with the second largest energy contribution.

optimization of one of its atoms followed by a full-dimensional local minimization) to be related, such that the latter is a "descendent" of the former. For some fixed number of iterations, the work in this phase is balanced over each of the k sets of configurations consisting of the k initial minimizers and all of those minimizer's descendants. First, each of the k initial minimizers is chosen for improvement twice, once using the atom that contributes the most to the total function value, and once with the atom that contributes the second most. Then, at each iteration for the remainder of the balancing phase, the set of configurations with the least amount of work performed on its members so far is selected, and the best configuration in this set that hasn't already been used twice is chosen. Configurations are rated in terms of their total energy function value, and best refers to the lowest in energy value. After the fixed number of iterations of the balancing phase have been performed, the remaining iterations of the local minimizer improvement phase select the best configuration that has not already been selected twice, regardless of where it is descended from. We have found that the combination of the breadth of search of the configuration space that the balancing phase provides with the depth of search that the non-balancing phase allows is useful to the success of our method.

The complete framework for the global optimization algorithm for solving Lennard-Jones problems is outlined in Algorithm 1 and Sub-algorithm 2. For further information on the algorithm, see [1].

To integrate smoothing techniques with Algorithm 1, the smoothed functions that will be used in the global optimization algorithm must be partially separable. Both our new smoothing function family (1.2) and the spatial-averaging function families in [6] [22] are partially separable. In part for this reason, the approach we describe now could be used with either type of smoothing.

One straightforward use of smoothing functions with Algorithm 1 would be to solve some smoothed problem $\mathbf{LJ}_s$, (equation (3.5)) by Algorithm 1, and then track the global minimizer or a small number of minimizers with the lowest energy values back to the original Lennard-Jones problem by the trajectory tracking procedure. As we have discussed, even if the global minimization algorithm on the smoothed function is successful, this approach could fail to find the global minimizer of the original problem, due to trajectory termination and/or order flips. If the unsmoothed global minimizer's trajectory terminates before smoothing level s, no trajectory in $\mathbf{LJ}_s$ will lead directly to the unsmoothed global minimum, and it is unlikely (but not impossible, due to jumps) that reverse tracking will locate the unsmoothed global minimizer. If the global minima's trajectory does not terminate but drops its order significantly, then the corresponding smoothed minimizer will not be in the small group of minimizers of $\mathbf{LJ}_s$ that is tracked back, and again the unsmoothed global minimizer will not be found. The design of the new algorithm should address and overcome these possible (and indeed, common) behaviors.

The design of the new algorithm that utilizes smoothing and the approach of Algorithm 1 is based on two observations. First, Phase 2 of Algorithm 1 can effectively find lower local minimizers that are structurally related to the current set of local minimizers. Secondly, even when the set of best smoothed minimizers does not include ones on the trajectories from best unsmoothed minimizers, due to the smoothing process it is very likely to include minimizers that are structurally related to the best unsmoothed minimizers.

The framework of the new algorithm is to solve the most smoothed problem $\mathbf{LJ}_{s_1}$ by Algorithm 1, track some number of the lowest minimizers for this problem to some less smoothing level s_2, and then use the best of these minimizers as starting points for Phase 2 of Algorithm 1 on this less smoothed problem. Then this process of tracking the best minimizers to less smoothed levels and applying Phase 2 is repeated for successively less smoothed functions, ending with the original unsmoothed problem. The algorithm depends on the ability of Phase 2 to eventually locate the global minimizer's trajectory, even if it is not present in the initial stages of the method. The overall sequence of smoothing parameter sets is denoted

$$(4.1) \qquad\qquad \mathbf{S} : \{s_1, s_2, \ldots, s_i, \ldots, s_{LJ}\}$$

where s_i is smoother than s_{i+1} and s_{LJ} represents Lennard-Jones potential. If our two-parameter function family (3.1) is used, each element of $\mathbf{S}$ is a pair, $s_i = (P_i, \gamma_i)$. The new algorithm is outlined in Algorithm 3.

As in Algorithm 1, the new algorithm begins with the sample generation phase, but now on the most smoothed function $\tilde{f}_{s_1}(x)$. After this, the algorithm can be viewed as a sequence of applications of Phase 2 of the original algorithm, each on a different, successively less smooth function $\tilde{f}_{s_i}(x)$. As in Algorithm 1, each application of Phase 2 iterates through steps 2a to 2e a number of times. The number of iterations within each Phase 2 may be predefined, and is an important parameter to the efficiency and effectiveness of the method; if too many iterations are taken at each application of Phase 2 the overall method may be inefficient, but enough iterations should be taken so that it is likely that the global minimizer of $\tilde{f}_{s_i}(x)$ has been found.

After performing Phase 2 at each smoothing level s_i except the last level (s_{LJ}), a new Phase 3 is performed. This phase transfers some number of the lowest minimizers of $\tilde{f}_{s_i}(x)$ that have been found to the next less smoothed level s_{i+1}, by trajectory tracking. After the transformations, the best local minimizers of $\tilde{f}_{s_{i+1}}(x)$ that the tracking process produces serve as the starting points for Phase 2 at smoothing level s_{i+1}. In practice, if two smoothing levels are not very far apart, one local minimization may suffice to track each minimizer at smoothing level s_i to smoothing level s_{i+1}. If this procedure is used and the number of minimizers that is transferred in Phase 3 is limited, the cost of Phase 3 will be small relative to the cost of Phase 2.

There are three important parameter choices in Algorithm 3. These are the smoothing level to start with, the number of smoothing stages, and the number of iterations within Phase 2 at each stage. In general, we want to start from a smoothing level where the number of minimizers has been significantly reduced, but the problem is still not too dissimilar from the original problem in its coarse grain structure. Also, we do not want the number of smoothing stages or the number of iterations within each Phase 2 to be too large, for efficiency reasons. On the other hand, if we make the jumps in smoothing level too large or the amount of work within Phase 2 too small, we may not solve the problem. Proper choices for these parameters appears to be mainly a matter of experience. We expect that as the difficulty of the problem increases, either in the number of parameters, the number of low minimizers, or the sizes of the basins of attraction of the lowest minimizers, more smoothing levels and more work in each Phase 2 will be required. It is important to note that it not necessary to start the new algorithm from a smoothing level where only few minimizers are left, since Phase 2 of the algorithm allows known minimizers to be improved effectively.

At this time, we have considerable experience with running Phase 1 of Algorithm 3 followed by simple trajectory tracking, and only very limited experience with running Algorithm 3 in its entirety. The next section will just briefly describe our experience with running Phase 1 on smoothed

---

Algorithm 3 – Framework of the Large-Scale Global Optimization Algorithm with Smoothing Technique for Lennard-Jones Problems

- Select a sequence of smooth parameter sets,
  $\mathbf{S}: \{s_1, s_2, \ldots, s_i, \ldots, s_{LJ}\}$, where s_i is smoother than s_{i+1} and s_{LJ} represents Lennard-Jones potential.
- Select a sequence of integers, $\mathbf{I}: \{I_1, I_2, \ldots, I_i, \ldots, I_{LJ}\}$.

1. **Sample Generation Phase**
   a. **Sampling in Full Domain :** Randomly generate the co-ordinates of the sample points in the sampling domain, and evaluate $\tilde{f}_{s_1}(x)$ at each new sample point. Discard all sample points whose function value is above a global "cutoff level".
   b. **One-atom Sampling Improvement :** Same as Algorithm 1, step 1b.
   c. **Start Point Selection :** Same as Algorithm 1, step 1c.
   d. **Full-Dimensional Local Minimizations :** Same as Algorithm 1, step 1d.

2. **Local Minimizer Improvement Phase:**
   Repeat step 2a to 2e for I_i times using objective function $\tilde{f}_{s_i}(x)$.
   a. **Select a Configuration :** Same as Algorithm 1, step 2a.
   b. **Expansion :** Same as Algorithm 1, step 2b.
   c. **One Atom Global Optimization :** Same as Algorithm 1, step 2c.
   d. **Full-Dimensional Local Minimization :** Same as Algorithm 1, step 2d.
   e. **Merge the New Local Minimizers :** Same as Algorithm 1, step 2e.

3. **De-smooth Configurations Between Smooth Levels:**
   a. **Termination Checking :** If $i = LJ$, then stop the algorithm.
   b. **Minimizer Transformation :** Apply the local optimization by using $\tilde{f}_{s_{i+1}}(x)$ as objective function, or apply the trajectory tracking procedure from smoothing level s_i to s_{i+1}, on some number of best configurations in the existing list of local minimizers. Increase i by one, and return to phase 2.

Lennard-Jones problems. It will show the advantages of the combination of smoothing and global optimization as is proposed in this section. In particular, it will be seen that applying Phase 1 to the smoothed problems does not necessarily locate the trajectory of the global minimizer, but it consistently locates the trajectories of far lower minimizers than are found by applying Phase 1 to the original unsmoothed function.

5. Numerical results. In this section we discuss an experiment that is a first step in assessing the effectiveness of the new algorithm presented in the previous section. This experiment gives a preliminary indication of the effectiveness of smoothing techniques integrated within a sophisticated global optimization algorithm. Its purpose is to examine the quality of the minimizers that are generated in Phase 1 and will be passed to Phase 2 of Algorithm 3 as starting points. Therefore, only Phase 1 of the new algorithm is utilized in the experiment.

We have applied Phase 1 of Algorithm 1 and of Algorithm 3 on Lennard-Jones problems with 20, 25, 30, 35, 40, 45, 50, 55, and 60 atoms. Phase 1 of Algorithm 3 was used with smoothing level (P=4.0,$\gamma = 0.2$) in these experiments. Each Phase 1 run was targeted to generate 93 minimizers. However, some minimizers are found more than once when Phase 1 is used on the smoothed functions. That is, the new Phase 1 generates fewer distinct minimizers. This is an expected consequence of using a smoothed objective function.

To assess the quality of the minimizers generated in Phase 1 on the smoothed problem, we have tracked each minimizer back to the unsmoothed problem. In our global optimization algorithms, we generally pass only the best 20 minimizers from Phase 1 on to Phase 2. Therefore in the results we report, we show the best unsmoothed minimizer that results from tracking back the 20 best smoothed minimizers. We also show the best minimizer that results from tracking back all the smoothed minimizers (up to 93) that were generated in Phase 1.

The results of these experiments are listed in Table 5.1. The second column of Table 5.1 lists what is currently believed to be the global minimum energy for the Lennard-Jones problem with this number of atoms. The third column shows the best value found in Phase 1 of Algorithm 1, i.e. on the unsmoothed problem, out of 93 generated minimizers. The fourth column shows the best Lennard-Jones value found by tracking the best 20 smooth minimizers found in Phase 1 of Algorithm 3, i.e. Phase 1 on the smoothed problem. The fifth column shows the best Lennard-Jones value found by tracking all the minimizers found in Phase 1 of Algorithm 3, as well as, (in parentheses) the order of the smooth minimizer leading to the unsmoothed global minimizer. The last column lists the numbers of distinct minimizers found in Phase 1 of Algorithm 1 / Algorithm 3, out of a possible 93.

From Table 5.1, we can see that Phase 1 applied to the smoothed

TABLE 5.1

Performance comparison between Phase 1 of Algorithm 1 and Phase 1 of Algorithm 3.
The smoothing level which is used in Phase 1 of Algorithm 3 is $(P = 4.0, \gamma = 0.2)$.

| # of Atoms | Best Known LJ Value | Best Found by Phase 1 on: | | | # of Minima Found |
| | | Alg. 1 LJ Potential | Alg. 3 Smoothed Potential | | Alg.1/Alg.3 |
			Track Best 20	Track All*	
20	-77.1770	-77.17704	-77.17704	-77.17704(1)	89/22
25	-102.3727	-101.8780	-102.3727	-102.3727(1)	93/30
30	-128.2866	-126.9128	-128.1816	-128.1816(1)	93/53
35	-155.7566	-153.4610	-155.7566	-155.7566(20)	93/58
40	-185.2498	-184.4230	-181.5001	-185.2498(38)	93/70
45	-213.7849	-208.8027	-209.0986	-211.3767(40)	93/73
50	-244.5499	-236.7978	-243.9400	-244.5499(53)	93/68
55	-279.2485	-269.5153	-274.3111	-276.6043(58)	93/70
60	-305.8755	-295.4530	-298.0514	-302.4619(60)	93/75

*:the number in parentheses is the order of the smooth minimizer which leads to the best minimizer in Lennard-Jones potential through tracking.

Lennard-Jones energy function consistently locates minimizers that track back to lower unsmoothed minimizers than the minimizers found by applying Phase 1 directly to the unsmoothed problem. In all the cases tried beyond 20 atoms, the lowest function value of all the minimizers found using Phase 1 on the smoothed problem and tracking is significantly closer to the putative global minimum value than the lowest minimizer found in Phase 1 on the unsmoothed problem, generally by a factor at least 3. Indeed, just applying Phase 1 to the smoothed function, followed by tracking, locates the global minimizers for the 20, 25, 35, 40, and 50 atom problems, whereas applying Phase 1 to the unsmoothed problem only locates the global minimizer for 20 atoms among the cases tried.

An interesting phenomenon shown by Table 5.1 is that for the problems with 40, 45, 50, 55, and 60 atoms, the lowest function value found by tracking all the smoothed minimizers is lower than the lowest found by tracking just the 20 best smoothed minimizers. Sometimes the difference is quite large, including in the 40 and 60 atom case, where the unsmoothed global minimum comes from a smoothed minimizer whose order is beyond the top twenty. This data indicates that for the larger problems, a large amount of order flipping is occurring at this smoothing level. It also suggests that in Algorithm 3, it might be fruitful to transfer considerably more than 20 smoothed minimizers from Phase 1 to the initial Phase 2. Note that the 40 atom problem is the sole case where the lowest function value found by tracking just the top 20 smoothed global minimizers is higher than the lowest minimizer found in Phase 1 on the unsmoothed problem.

Another observation from Table 5.1 is that Phase 1 on the smoothed objective function consistently found fewer *distinct* minimizers than Phase 1 on the unsmoothed problem. This is an indication that the number of minimizers in the smoothed Lennard-Jones problems has been significantly reduced in comparison to the original problem. Due to the heuristic nature of our algorithm and the different molecular structure of different size clusters, we can not observe a clear trend in the number of distinct minimizers found by the new Phase 1. However, in general, larger smoothed problems appear to have more distinct minimizers than smaller smoothed problems. This indicates that, as expected, the numbers of minimizers in smoothed Lennard-Jones problems at the same smoothing level increases as the number of atoms increases.

In summary, Table 5.1 clearly shows that the quality of the minimizers found in Phase 1 of our global optimization method has been significantly improved by integrating the algorithm with smoothing techniques. Since the Phase 1 minimizers serve as input to Phase 2 of the global optimization algorithm, their quality is an important factor in the performance of the whole global optimization algorithm. The results of this section make in seem likely that Algorithm 3 should be able to find the global minimizer for these problems considerably more efficiently than Algorithm 1 does. This hypothesis can only be verified through careful experimentation with the full algorithm, however. These experiments are currently in progress and their results will be reported in a subsequent paper.

6. Summary and future research. The first part of this paper suggests a new family of functions for smoothing the Lennard-Jones potential energy function. The smoothing functions naturally remove the pole from the Lennard-Jones potential and widen the basin of attraction of its minimizer, without changing the location or energy value of the minimizer. We show the similarity between the new family of smoothing functions and the spatial-averaging function family in [6] [22]. However, the two smoothing parameters of the new function family appear to provide a richer choice of smoothed function.

The first set of experiments in this paper show that the trajectories of local minimizers generated through the new smoothing function family have some potentially undesirable behaviors. The undesired behaviors, which almost certainly are shared by other families of smoothing functions, are order flips, termination of trajectories, and emergence of new trajectories. These behaviors show that an algorithm that simply finds the global minimizer of a very smoothed problem and tracks it (or possibly the lowest few minimizers) back to the unsmoothed problem cannot be expected to locate the global minimizer of the unsmoothed problem in general. However, the experiments do demonstrate that the new smoothing function family effectively reduces the number of minima in the Lennard-Jones problem. In addition, the experiments indicate that the lowest minimizers of the smoothed problem may track back to relatively low minimizers of the unsmoothed problem.

The next part of this paper describes a new global optimization algorithm that is intended to integrate smoothing and sophisticated global optimization in a way that overcomes the possible undesirable behavior of smoothing trajectories. This new algorithm is based largely upon the global optimization algorithm in [1]. The new algorithm iterates through a sequence of selected smoothing levels that gradually de-smooth the problem back to the original one. Phase 1 of the algorithm generates an initial set of low minimizers at the smoothest level and passes these starting points to Phase 2. Phase 2 is performed at each smoothing level. As in the algorithm in [1], Phase 2 performs a sequence of iterations that attempt to find lower local minimizers than the ones it is given initially. The main work of this phase is conducted through a set of carefully selected small-scale global optimization problems. This phase has the ability to discover lower minimizers that are not on the trajectories that have been discovered so far. With this ability, the algorithm can overcome the possible undesirable behaviors of the smoothing trajectories. After each Phase 2 except the last (i.e. on the unsmoothed problem), a Phase 3 is performed that transfers the best few minimizers that Phase 2 has found at the currenting smoothing level to next, less-smoothed level. This is done either by a simple local optimization or a trajectory tracking procedure, depending on the distance between the smoothing parameters from one level to the next. After Phase 3 is completed, the algorithm goes back to Phase 2 at the next smoothing level, using the minimizers just generated in Phase 3 as the initial set for Phase 2. The algorithm keeps iterating through Phase 2 and Phase 3 until the function has been de-smoothed to the original problem.

The hope in combining smoothing with a sophisticated global optimization algorithm in this way is that even if the smoothed minimizers may not lie on trajectories leading to the best unsmoothed minimizer, they will lie on trajectories leading to fairly good unsmoothed minimizers. Furthermore, it is hoped that these good unsmoothed minimizers will have some degree of structural similarly to the best unsmoothed minimizers. Therefore, Phase 2 of the global optimization algorithm, which is particularly good at discovering lower local minimizers that are structurally similar to known local minimizers, should be effective in finding the best minimizers from the smoothed minimizers.

The final portion of this paper reports on an experiment that is intended to be a preliminary assessment of whether these hopes are realized. It examines whether Phase 1 of the algorithm of [1] (which is also the initial phase of the new algorithm) generates better minimizers when applied to the smoothed problem than when applied to the unsmoothed problem. The goodness of the smoothed minimizers is assessed by tracking them back to the unsmoothed problem. The results of the experiment show a fairly dramatic advantage for Phase 1 applied to the smoothed problem as compared to Phase 1 applied to the unsmoothed problem. These results give reason to expect that the new global optimization algorithm will solve Lennard-Jones problems more efficiently and effectively than the algorithm without smoothing.

The immediate next step of this research is to completely test the new global optimization / smoothing method on Lennard-Jones problems. Experiments that explore the effectiveness of the new algorithm are currently underway. The results will be reported in a forthcoming paper. There are a variety of algorithmic options that need to be explored in the new method. These include how far the function should be smoothed initially, how many smoothing stages should be used, how much effort should be allocated at each stage, and how many minimizers should be transferred between stages. The results of the current experiments should provide some insight into these issues as well as about the overall effectiveness of the new approach for Lennard-Jones problems.

The larger goal of this research is to apply the types of techniques described in this paper to more complex molecular conformation problems, including the protein folding problem. The algebraic smoothing technique described in this paper can be generalized to other functions (usually with poles) that are likely to benefit from smoothing, such as electrostatic forces or attractive/repulsive forces given by different formulas than the Lennard-Jones potential. Once a family of smoothing functions is given, the global optimization / smoothing algorithm that is proposed in this paper can readily be generalized to apply to a wide range of molecular configuration problems. Very recently, we have performed some preliminary experiments on the protein polyalanine where we smooth the Lennard-Jones and electrostatic terms in the potential by techniques related to those in this paper (leaving the remaining, less troublesome terms in the potential unchanged) and then perform a different Phase 1 process described in [7] followed by trajectory tracking. The minimizers obtained by this approach so far are vastly superior to those obtained by Phase 1 without tracking. This very preliminary work indicates that the techniques suggested in this paper may have promise for more complex and important molecular conformation problems. The results of this research also will be reported in future papers.

REFERENCES

[1] BYRD, R.H., ESKOW, E., AND SCHNABEL, R.B., *A New Large-Scale Global Optimization Method and its Application to Lennard-Jones Problems*. Technical Report CU-CS-630-92, Dept. of Computer Science, University of Colorado, revised, 1995.

[2] BYRD, R.H., DERBY, T., ESKOWM, E., OLDENKAMP, K., AND SCHNABEL, R.B. *A New Stochastic/Perturbation Method for Large-Scale Global Optimization and its Application to Water Cluster Problems*, in Large-Scale Optimization: State of the Art, W. Hager, D. Hearn, and P. Pardalos, eds., Kluwer Academic Publishers, Dordrecht, The Netherlands, pp. 71–84, 1994.

[3] BYRD, R.H., DERT, C.L., RINNOOY KAN, A.H.G., AND SCHNABEL, R.B *Concurrent stochastic methods for global optimization, Mathematical Programming*, **46**, 1–29, 1990.

[4] COLEMAN, T., SHALLOWAY, D. AND WU, Z. *Isotropic Effective Energy Simulated Annealing Searches for Low Energy Molecular Cluster States*, Technical Report CTC-92-TR113, Center for Theory and Simulation in Science and Engineering, Cornell University, 1992.

[5] COLEMAN, T., SHALLOWAY, D. AND WU, Z. *A Parallel Build-up Algorithm for Global Energy Minimizations of Molecular Clusters Using Effective Energy Simulated Annealing*, Technical Report CTC-93-TR130, Center for Theory and Simulation in Science and Engineering, Cornell University, 1993.

[6] COLEMAN, T., AND WU, Z. *Parallel Continuation-Based Global Optimization for Molecular Conformation and Protein Folding*, Technical Report CTC-94-TR175, Center for Theory and Simulation in Science and Engineering, Cornell University, 1994.

[7] BYRD, R.H., ESKOW, E., VAN DER HOEK, A.,SCHNABEL, R.B, SHAO, C-S., AND ZOU, Z. *Global optimization methods for protein folding problems*, to appear in Proceedings of the DIMACS Workshop on Global Minimization of Nonconvex Energy Functions: Molecular Conformation and Protein Folding, P. Pardalos, D. Shalloway, and G. Xue, eds., American Mathematical Society, 1995.

[8] FARGES, J., DEFERAUDY, M.F., RAOULT, B., AND TORCHET, G. *Cluster models made of double icosahedron units, Surface Sci.*, **156**, 370–378, 1985.

[9] FREEMAN, D.L. AND DOLL, J.D. *Quantum Monte Carlo study of the thermodynamic properties of argon clusters: the homogeneous nucleation of argon in argon vapor and "magic number" distributions in argon vapor, J. Chem. Phys.*, **82**, 462–471, 1985.

[10] HOARE, M.R. *Structure and dynamics of simple microclusters, Adv. Chem. Phys.*, **40**, 49–135, 1979.

[11] HOARE, M.R. AND PAL, P. *Physical cluster mechanics: statics and energy surfaces for monatomic systems, Adv. Phys.*, **20**, 161–196, 1971.

[12] JUDSON, R.S., COLVIN, M.E.I, MEZA, J.C., HUFFER, A. AND GUTIERREZ, D. *Do Intelligent Configuration Search Techniques Outperform Random Search for Large Molecules?*, SAND91-8740, 1991.

[13] KOSTROWICKI, J., PIELA, L., CHERAYIL, B.J., AND SCHERAGA, A. *Performance of the Diffusion Equation Method in Searches for Optimum Structures of Clusters of Lennard-Jones Atoms, J. Phys. Chem.*, **95**, 4113–4119, 1991.

[14] MARANAS, C. D. AND FLOUDAS C. A. *A Global Optimization Approach for Lennard-Jones Microclusters*, Preprint, Department of Chemical Engineering, Princeton University, June 1992.

[15] NORTHBY, J.A. *Structure and binding of Lennard-Jones clusters:$13 \leq N \leq 147$, J. Chem. Phys.*, **87**, 6166–6177, 1987.

[16] ORESIC, M., AND SHALLOWAY, D. *Hierarchical Characterization of Energy Landscapes using Gaussian Packet States, J. Chem. Phys.*, **101**, 9844–9857, 1994.

[17] PILLARDY, J., AND PIELA, L. *Molecular Dynamics on Deformed Potential Energy Hypersurfaces, J. Phys. Chem.*, **99**, 11805–11812, 1995.

[18] SCHNABEL, R.B., KOONTZ, J.E., AND WEISS, B.E. *A modular system of algorithms of unconstrained minimization, ACM Transactions on Mathematical Software*, **11**, 419–440, 1985.

[19] SHALLOWAY, D. *Packet annealing : a deterministic method for global minimization, with application to molecular conformation. Preprint, Section of Biochemistry, Molecular and Cell Biology, Global Optimization*, C. Floudas and P. Pardalos, eds., Princeton University Press, 1991.

[20] SMITH, S.L., ESKOW, E., AND SCHNABEL, R.B. *Adaptive, asynchronous stochastic global optimization algorithms for sequential and parallel computation, In Proceedings of the Workshop on Large–Scale Numerical Optimization, T.F.*

Coleman and Y. Li, eds. SIAM, Philadelphia, 207–227, 1989.

[21] WILLE, L.T. *Minimum-energy configurations of atomic clusters: new results obtained by simulated annealing*, Chem. Phys. Lett., **133**, 405–410, 1975.

[22] WU, Z. *The Effective Energy Transformation Scheme as a Special Continuation Approach to Global Optimization with Application to Molecular Conformation*, Technical Report CTC-93-TR143, Center for Theory and Simulation in Science and Engineering, Cornell University, 1993.

[23] XUE, G. *Improvement of the Northby Algorithm for Molecular Conformation: Better Solutions,* , J. Global Optimization, **4**, 425. 1994.

IMA SUMMER PROGRAMS

1987 Robotics
1988 Signal Processing
1989 Robustness, Diagnostics, Computing and Graphics in Statistics
1990 Radar and Sonar (June 18 - June 29)
      New Directions in Time Series Analysis (July 2 - July 27)
1991 Semiconductors
1992 Environmental Studies: Mathematical, Computational, and
      Statistical Analysis
1993 Modeling, Mesh Generation, and Adaptive Numerical Methods
      for Partial Differential Equations
1994 Molecular Biology
1995 Large-Scale Optimizations with Applications to Inverse Problems,
      Optimal Control and Design, and Molecular and Structural
      Optimization
1996 Emerging Applications of Number Theory
1997 Statistics in Health Sciences
1998 Coding and Cryptography

SPRINGER LECTURE NOTES FROM THE IMA:

The Mathematics and Physics of Disordered Media
    Editors: Barry Hughes and Barry Ninham
    (Lecture Notes in Math., Volume 1035, 1983)

Orienting Polymers
    Editor: J.L. Ericksen
    (Lecture Notes in Math., Volume 1063, 1984)

New Perspectives in Thermodynamics
    Editor: James Serrin
    (Springer-Verlag, 1986)

Models of Economic Dynamics
    Editor: Hugo Sonnenschein
    (Lecture Notes in Econ., Volume 264, 1986)

The IMA Volumes in Mathematics and its Applications

Current Volumes: